S0-BRA-485

Parallel Robotic Machine Tools

Dan Zhang

Parallel Robotic Machine Tools

Dan Zhang
Faculty of Engineering and Applied Science
University of Ontario Institute of Technology (UOIT)
Oshawa, ON
Canada
dan.zhang@uoit.ca

ISBN 978-1-4419-1116-2 e-ISBN 978-1-4419-1117-9
DOI 10.1007/978-1-4419-1117-9
Springer New York Dordrecht Heidelberg London

Library of Congress Control Number: 2009939938

Printed on acid-free paper

Springer is part of Springer Science+Business Media (www.springer.com)

Dedicated to my parents: Hongtao Zhang, Yu Li who have inspired the thirst for knowledge in me all the time.

To my wife, Junmei Guo, daughter, Mengjia Zhang and son, James William Zhang for all their support.

Preface

Scope of the Book

Research and development of various parallel mechanism applications in engineering is now being performed in every industrial field. Parallel robot-based machine tool development is considered a key technology of robot applications in the future of manufacturing industries. The study described in this book is concerned with the basic theory, approaches, and algorithms in the field of parallel robot-based machine tools. A family of new alternative mechanical architectures which could be used for machine tools with parallel architecture is introduced. The kinematic analysis, stiffness analysis, kinetostatic modeling, optimization, design of these mechanism systems, and reconfigurable parallel kinematic machine tools are also discussed in the book.

The book includes the basic conceptions in parallel kinematic machines at the forefront of this field. It can be used as graduate textbook in advanced machine tools, or as a research monograph. This book is also suitable as a reference for engineers, researchers, and students who range from senior undergraduates to doctoral students who are interested in parallel robotics or advanced machine tools technology. This book gives the audience a deep understanding of the classical applications of parallel mechanisms in the field of mechanical manufacturing.

Features of the Book

- This book focuses on the junction of parallel robot and machine tools. A successful application of parallel mechanisms in the field of machine tools, which is also called parallel kinematic machines tools, has been the developing trend of advanced machine tools.
- This book results from the author's research in the field of parallel robotic machine tools over the last 10 years.
- This book not only includes the main aspects and important issues of robotic machine tools, but also references novel conceptions and approaches, i.e. the general kinetostatic model, artificial intelligence-based performance optimization, global stiffness model, and others.

- Most of the existing books regarding parallel kinematic machines were built upon the concept of traditional "Gough–Stewart" mechanism types. This suggests that most of the parallel mechanisms developed have six degrees of freedom. However, in many applications, five or less degrees of freedom are required. Hence, there is a need to study parallel mechanisms with less than six degrees of freedom, which this book focuses on.
- The analysis approaches proposed in this book are novel. If audiences can understand these issues and grasp these analytic approaches, they will open the gate to the advancing fronts of parallel robot-based machine tools.
- Other related books are complex and hard to read. As the organization principles of this book are from easy to hard, audiences can easily access the keys to understanding the theories of this book.
- A large number of case studies and numerical analyses help the audience master the main ideas of the book in both theory and practice.

Oshawa, Ontario, Canada, June 2009 Dan Zhang

Contents

Chapter 1
Introduction

1.1 The History of Parallel Robots

The demands for higher performance of general-purpose industrial robots are increasing continuously. In particular, the need for truly adaptive automation in many applications has led to higher requirements for operational accuracy, load capacity, task flexibility, reliability, and cycle time with robots. Examples of such needs are higher precision assembly, faster product handling, better measurements, surface finishing, and milling capabilities. Furthermore, there is a high demand for off-line programming to eliminate touch-up of programmed positions; in other words, robots must perform their task with better load capacity and accuracy in operations. A general trend of meeting these demands and requirements is to make use of parallel robots, which have excellent potential capabilities, including high rigidity, high accuracy, and high loading capacities.

Parallel robots generally comprise two platforms, which are connected by at least two kinematic chains, and to provide relative motion between a moveable platform and a base platform. In fact, parallel robots have become an indispensable part of general robots both in industry and in academia. Besides, with the rapid development of parallel robots a few decades ago, the research on mechanism theory, mobility analysis, dimensional synthesis, kinematics and dynamics modeling, and design optimization have been increasing in a large scale.

Centuries ago, the English and French mathematicians attained a keen interest in polyhedral. It was from this obsession that the first theoretical works involving parallel mechanisms, specifically six-strut platforms, were developed. However, there were very few scholars who actually read and studied these works.

In 1928, a spherical parallel robot (shown in Fig. 1.1) as a conceptual amusement device was invented by James E. Gwinnet. This is perhaps the first spatial parallel mechanism. Unfortunately, the entertainment industry did not pay attention to such an invention at that time.

Ten years later, Willard L.V. Pollard designed a novel parallel robot for automatic spray painting, which was claimed as the first industrial parallel robot. This three-legged robot was capable of five degree-of-freedom motion – three for the position of the tool head, and the other two for orientation. However, this robot was

D. Zhang, *Parallel Robotic Machine Tools*, DOI 10.1007/978-1-4419-1117-9_1,

Fig. 1.1 Potentially the first spatial parallel mechanism, patented in 1931 (US Patent No. 1,789,680)

never actually built and Pollard's son, Willard L.V. Pollard Jr. actually designed and engineered the first industrial parallel robot, as shown in Fig. 1.2.

In 1947, Dr. Eric Gough invented a new six degree-of-freedom parallel robot that revolutionized the robotic industry – the first octahedral hexapod (called the universal rig by him). It was applied as a tire-testing apparatus (see Fig. 1.3 (left)) to discover properties of tires subjected to different loads. Figure 1.3 (right) shows the machine which was put into use in 1954 and retired in 2000. This platform consists of six identical extensible links, connecting the fixed base to a moving platform to which a tire is attached. The kinematic chains associated with the six legs, from base to platform, consist of a fixed Hooke joint, a moving link, an actuated prismatic joint, a second moving link and a ball-and-socket joint attached to the moving platform. The position and the orientation of the moving platform, together with the attached wheel, are changed according to the variation of the links length. This wheel is driven by a conveyor belt, and the mechanism allows the operator to measure the tire wear and tear under various conditions. The universal rig has been playing an important role in the field of industry robots and still has great effect for the academic research of parallel manipulators. Many significant advantages can be discovered when compared with conventional serial counterparts, such as higher stiffness and payload, higher force/torque capacity, lower inertia, eminent dynamic characteristics, less accumulated error of joints, and parallel robots also have simpler inverse kinematics which is convenient for real-time control. Some disadvantages also should be mentioned, such as smaller workspace, worse dexterity.

In 1965, Stewart published a paper describing a 6DOF motion platform that was designed as an aircraft simulator. The so-called "Stewart platform" was a parallel mechanism that differentiated from the octahedral hexapod. Figure 1.4 is a schematic of the Stewart platform. Stewart's work had a significant impact on the further development of parallel mechanisms in which he made many suggestions

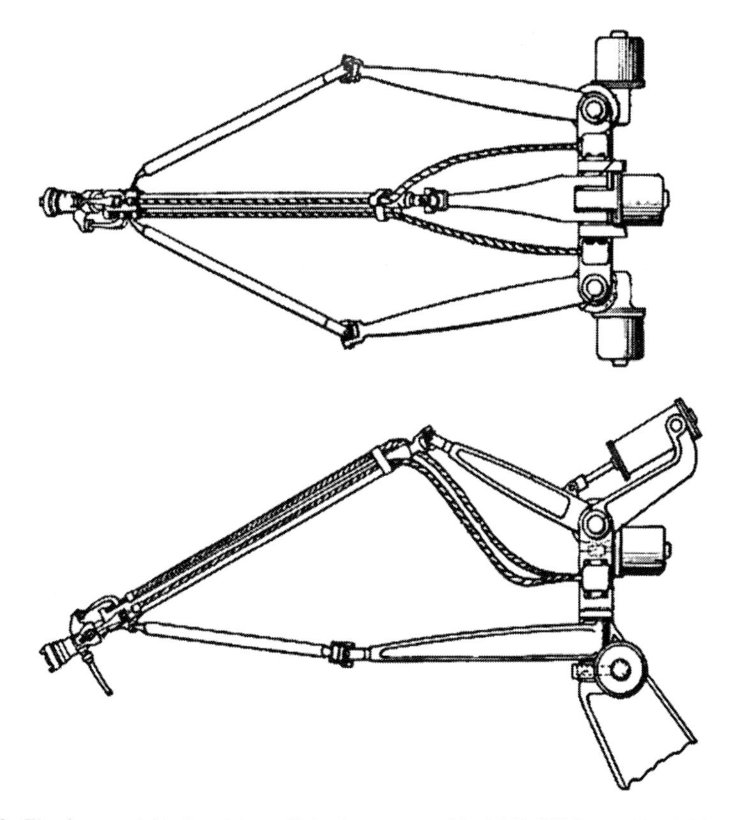

Fig. 1.2 The first spatial industrial parallel robot, patented in 1942 (US Patent No. 2,286,571)

Fig. 1.3 The first octahedral hexapod (left, original Gough platform) developed in 1954; and the Gough platform for tire test (right)

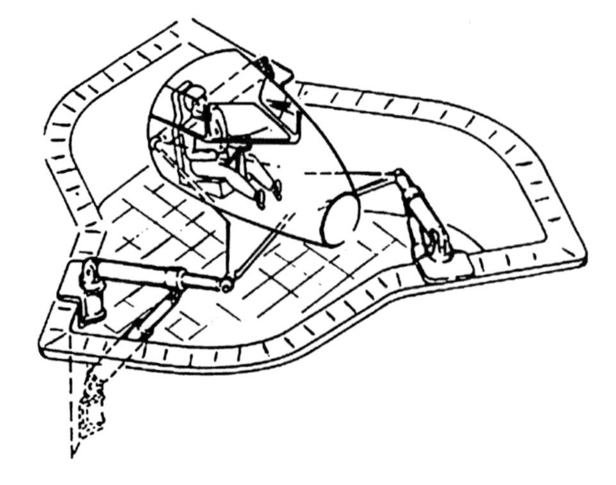

Fig. 1.4 Stewart Platform [137]

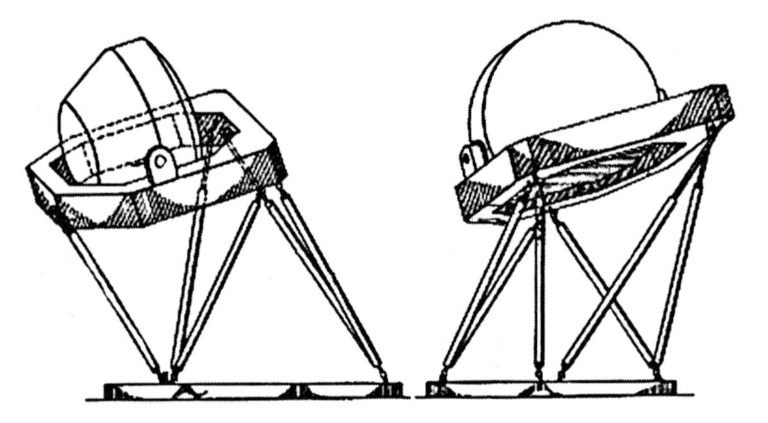

Fig. 1.5 An octahedral hexapod patent issued in 1967 (US patent No. 3,295,224)

for uses of the hexapod, and which eventually became reality. He was also responsible for popularizing Gough's design in academia. In fact, the contribution of Gough established the milestone for the development of parallel robots in industry, while, it is Stewart who introduced it to academia. Over the past decades, there were many new mechanisms that had been proposed and released by researchers, anyhow, not so many are adopted by industry.

It is noticed that, in 1962, an engineer named Klaus Cappel, who was from the Franklin Institute Research Laboratories in Philadelphia, proposed the same octahedral hexapod as Gough's, to be used as a motion simulator(as shown in Fig. 1.5). Cappel was granted a patent for his invention in 1967. He is considered as the third and last pioneer in the field of parallel robots.

Fig. 1.6 The first flight simulator based on an octahedral hexapod as in the mid 1960s (courtesy of Klaus Cappel)

The very first flight simulator based upon Cappel's design was constructed (as shown in Fig. 1.6). Cappel has also designed various parallel robot systems for vibration testing. However, it took a long time before these designs were accepted by industry.

It was these three men (Eric Gough, D. Stewart, and Klaus Cappel) who were truly the pioneers of the parallel robot. This has paved a way for many new inventions and applications of parallel mechanisms.

Nowadays, parallel robots can be found in many practical applications, such as aircraft and vehicle simulators [7, 11, 73, 74, 111, 114], adjustable articulated trusses [35, 65, 66, 139, 161], medical devices [27, 28, 97, 113, 130, 134, 153], micro-robot [40,41,78,119,162,163], and force/torque sensor [47,125,129,135]. More recently, they have been used in the development of high precision machine tools [15,92,152, 167] by many companies such as Giddings & Lewis, Ingersoll, Hexel, Geodetic and Toyoda, and others. The Hexapod machine tool [12, 15, 70, 104, 122] is one of the successful applications.

1.2 Introduction of Conventional Machine Tools

Machine tools are the fundamental implements that change the shape, surface, or properties of a blank object made of metal, plastic, wood, or other material. Producers' machine portfolios traditionally focus on a single process technology. Before the advent of numerical control technology, some producers specialized in turning

Fig. 1.7 The standard engine lathe (South Bend Lathe Corp)

machines, others in grinding machines, presses, etc. These machine tools include the following:

1. The lathe, or the engine lathe (Fig. 1.7), is used to produce round work. The workpiece, held by a work-holding device mounted on the lathe spindle, is revolved against a cutting tool, which produces a cylindrical form. Straight and taper turning, facing, drilling, boring, reaming, and thread cutting are some of the common operations performed on a lathe.
2. The drill (Fig. 1.8), which is one of the most common machine tools. Drills cut cylindrical holes in objects.
3. Milling Machine: The horizontal milling machine (Fig. 1.9), and the vertical milling machine (Fig. 1.10), are the two of the most useful and versatile machine tools. Both machines use one or more rotating milling cutters having single or multiple cutting edges.
4. Grinder (Fig. 1.11): grinding wheels remove material from an object by rubbing against it, slowly wearing the material away. The grinding wheel is typically used near the end of the machining task, to smooth a finish or to obtain extremely accurate dimensions.
5. Metal Saw: Sawing is a cutting operation in which the cutting tool is a blade (saw) having a series of small teeth, with each tooth removing a small amount of material. Metal-cutting saws are used to cut metal to the proper length and shape.

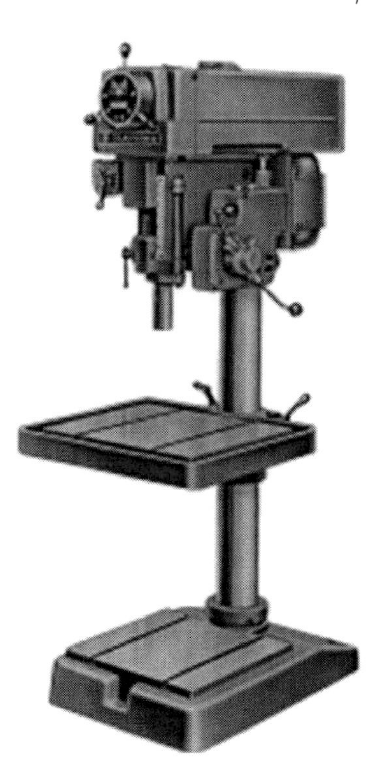

Fig. 1.8 The drill press is commonly used to produce holes (Clausing Industrial Inc)

Fig. 1.9 The universal knee and column type horizontal milling machine is used for machine flat, angular, and contoured surfaces (Cincinnati Machine, a UNOVA Co.)

Fig. 1.10 A standard vertical milling machine (Bridgeport Machines, Inc.)

Fig. 1.11 The surface grinder is commonly used to finish flat surfaces (DoAll Co.)

Machine tools have advanced along with technology. Inventions such as numerical control, computer numerical control, artificial intelligence, vision systems, superabrasive cutting tools, stereo lithography, etc., have changed the way goods are manufactured. These developments have improved machine tools and forever changed manufacturing processes, so that today it is possible to automatically produce high-quality products quickly, accurately, and at lower cost than ever before, and at the same time, the mechanism is also very different from that of the conventional machine tool. Modern machine tools based on parallel kinematic (PK) technology that offer faster, more accurate and less costly alternatives to conventional systems than most conventional machine tools for component manufacture or assembly. Conventional machine tools have serial knematic archtechture, with each axial of movement supporting the following axis and providing its motion. A significant drawback of conventional machine is that the moving parts must be heavy enough to provide the necessary stiffness to control the bending movements. This impacts the dynamic performance and reduces operational flexibility.

1.3 Parallel Robot-based Machine Tools

Because of the recent trend toward high-speed machining, there is a demand to develop machine tools with high dynamic performance, improved stiffness, and reduced moving mass. Parallel mechanisms have been adopted to develop this type of machine. Generally, parallel robot-based machine tool is called parallel kinematic machine [21, 175–177].

Hexapod machine tool, as one kind of parallel kinematic machines, has been widely studied and developed by researchers. Matar [104] defines a "Hexapod" as a geometric structure where a hexagon provides the points on a frame for six struts, which are then collected into pairs to form a triangle, whose position in free space can be uniquely described by the struts length.

The parallel kinematic mechanism offers higher stiffness, lower moving mass, higher acceleration, potential higher accuracy, reduced installation requirements, and mechanical simplicity relative to existing conventional machine tools [24, 127, 154, 160, 171]. By virtue of these attributes, the parallel kinematic mechanism offers the potential to change the current manufacturing paradigm. It has the potential to be a highly modular, highly reconfigurable, and high precision machine. Other potential advantages include high dexterity, the requirement for simpler and fewer fixtures, multi-mode manufacturing capability, and a small foot print. A comparison between the Hexapod machine tools and the conventional machine tools is given by Giddings and Lewis. It shows that the Hexapod machine tool has improved machine tools substantially in terms of precision (about 7 times), rigidity (about 5 times), and speed (about 4 times) [96].

So far, there are several companies and institutes involved in research and development of this kind of machine tool. Aronson [12] summarized the four major companies, and they are Giddings and Lewis, Ingersoll Milling Machine Co., Hexel

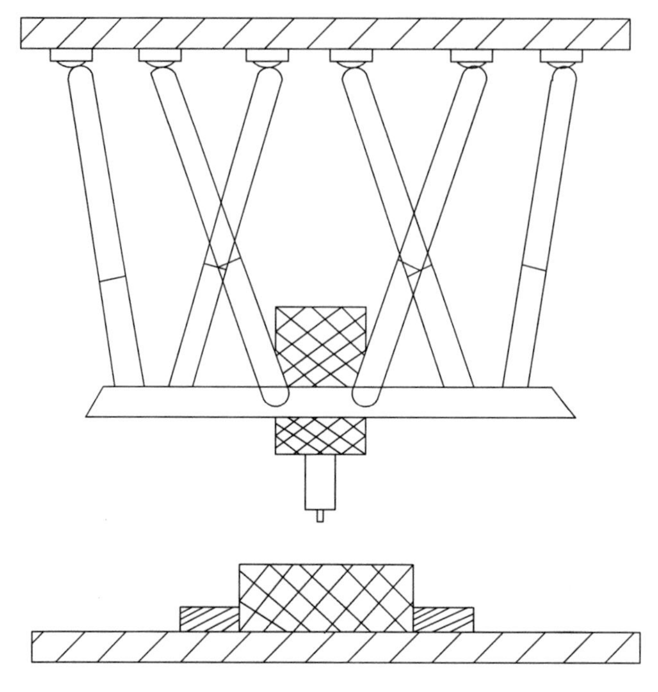

Fig. 1.12 Six-axis Hexapod machining center

Corporation and Geodetic Technology International Ltd. Giddings and Lewis did some of the early pioneering effort on the Variax, the Giddings and Lewis hexapod machine. Moreover, the industrial interest is continually growing [112].

Figure 1.12 represents a parallel mechanism module from Ingersoll Milling Machine Co. [93], it consists of a fixed upper dome platform and a moving lower platform, connected by six struts, which are precision ballscrews. On the upper platform, the six struts are driven by motor driven ballnuts. These alter the position and attitude of the lower platform by extending or retracting the struts. The ballscrews join the lower platform at three points, with two struts sharing a ball-and-socket joint. Various head attachments can be incorporated to suit a variety of applications. Each individual axis (leg drive) is independent from the others and comes with a personality file containing information such as error mapping (e.g., lead pitch variation), mounting offsets, physical performance, and thermal expansion characteristics. There are some other institutes and industry doing research and development work in this area. They are NEOS Robotics (Tricept series), Toyoda Machine Works (HexaM Machine), ITIA-CNR (ACROBAT), Seoul National University (ECLIPSE), Sandia Hexapod Testbed, Swiss Federal Institute of Technology (Hexaglide), Materials Engineering Division (MMED) from Lawrence Livermore National Laboratory (LLNL) (Octahedral Hexapod), SMARTCUTS (Simultaneous MAchining through Real Time Control of Universal Tooling System) (modular 3-DOF parallel link mechanism), LME Hexapod machine (Hexapod software model), University of Stuttgart (modular parallel mechanism design), and others.

Moreover, there are also many publications concerning the research and development of parallel kinematic machines. Heisel [68] presents the precision requirements for parallel kinematic machine tools design. Wang et al. [151] discuss the design and kinematics of parallel mechanisms for manufacturing. Pritschow and Wurst [122] describe a systematic design procedure that allows the evaluation of the technological feasibility of hexapods, and the parallel kinematic machines (PKMs) types that are currently being investigated by European researchers are presented in [121]. Wavering [155] introduces the history of PKMs research at the NIST Manufacturing Engineering laboratory, the current research areas and the potential directions for future work. Abbasi et al. [6] address a parametric design methodology for a special 6-6 parallel platform for contour milling. Warnecke et al. [154] present the analysis, designs, and variants of parallel-structure-based machine tools, different design variants are compared with regard to the load of the structures and the singularities. Gopalakrishnan and Kota [55] study various parallel manipulator configurations and the possibility of their integration under the evaluation of reconfigurable machining systems. The modular concepts for PKMs are proposed in the paper, similarly to [158]. An approach to Parallel Kinematic Machines design integrating tools for machine configuration, synthesis, and analysis is presented in [145]. Fassi et al. [45] present an approach to the development of a computer aided configuration tool for parallel kinematic machines. The goal of this tool is to enable a quick comparison between different machine structures. Bianchi et al. [22] propose a virtual prototyping environment for PKMs analysis to ease the industrial adoption of PKMs by availability of methodologies and integrated tools able to analyze PKMs of any architecture in a short period of time, providing the key data needed to design the machine. Weck et al. [156] discuss the substantial features of PKMs with special focus on structurally caused problems in design, control, and calibration and takes Ingersoll Octahedral Hexapod and the Dyna-M concept as examples for possible solutions. Some industrial applications are reported in the literature. For instance, Honegger et al. [71] present the adaptive control of the Hexaglide. Ryu et al. [132] present the "Eclipse" machine tool designed for rapid machining with their research of kinematic analysis. Powell et al. [120] focus on the Giddings and Lewis Variax Hexapod machine tool by presenting different metal cutting tests and analyzing the machine tools performances. Tönshoff et al. [144] present the structure and characteristics of the hybrid manipulator "Georg V" at Hannover University. Pierrot and Shibukawa [117] report the patented machine tools "HEXA" and "HexaM" at Toyoda Machine Works Ltd and Clavel [128] display the "Delta" parallel robot. Pritschow and Wurst [123] propose a systematic methodology for the design of different PKM topologies. Merlet [106] develops the software for the optimal design of a specific PKM class Stewart platform-based mechanisms. Boeij et al. [23] propose numerical integration and sequential quadratic programming method for optimization of a contactless electromagnetic planar 6-DOF actuator with manipulator on top of the floating platform. Chablat and

Angeles [31] investigate on optimum dimensioning of revolute-coupled planar manipulators based on the concept of distance of Jacobian matrix, to a given isotropic matrix which was used as a reference model. Zhang et al. [178] develope an integrated validation system for PKM that consists of kinematic/dynamic analysis module, kinetostatic model, CAD module, FEM module, CAM module, optimization module, and virtual environment for remote control. Pond and Carretero [118] apply the Jacobian matrix to determine the dexterity of parallel mechanisms regardless of the number and type of degrees of freedom of the mechanism. Company and Pierrot [38] develope a 3-axis PKM intended to be used for high-speed point-to-point displacement and simple machining. It was observed that the trajectory planning in the joint coordinate system is not reliable without taking into considerations of cavities or holes in the joint workspace. Li and Xu [98] study the stiffness characteristics of a three-prismatic-universal-universal translational PKM, where the stiffness matrix was derived intuitively with an alternative approach considering actuations and constraints. Bi and Lang [20] develope a concept so-called joint workspace for design optimization and control of a PKM, and some others.

In summary, all the existing parallel kinematic machines can be classified as follows:

1. From the viewpoint of the frame, two approaches to (PKMs) frame design exist. Ingersoll Milling Machine Co. (in conjunction with National Institute of Standards and Technology, NIST) (Fig. 1.12), Hexel Corporation, and Geodetic Technology International Ltd. all use a separate frame that suspends the hexapod, while Giddings and Lewis connected the spindle platform directly to the table platform (Fig. 1.13), thus avoiding thermal distortion and improving stiffness.
2. From the viewpoint of the structure, a new design called the Triax – not technically a hexapod – has been investigated by Giddings and Lewis. It will operate in only three axes. In contrast to the Hexapod machine from Ingersoll or Giddings and Lewis, The Institute for Control Technology of Machine Tools and Construction Units (ISW) of the University of Stuttgart has developed a Hexapod [122] whose motion is generated by linear movement of the base points of fixed length links and not by changing the leg length (Fig. 1.14). The Hexaglide [71] (Fig. 1.15) from Swiss Federal Institute of Technology also falls into this type.
3. From the viewpoint of workspace volume, the Hexaglide [71] (Fig. 1.15) from the Swiss Federal Institute of Technology differs from the Hexapod by the fact that the joints are placed on parallel guideways. Thus, instead of changing the total length of the legs, they have the possibility to make the guideways longer to extend the workspace of the machine in one direction. All other dimensions stay unaffected. This makes the Hexaglide an ideal mechanism for the machining of long parts. The Hexaglide is also easier to build and to measure than the Hexapod.
4. From the viewpoint of actuated joints, there are three types of parallel kinematic machines:

   - Prismatic actuated machines with variable leg lengths and fixed joints (e.g., Ingersoll, Neos Robotics),

Fig. 1.13 The Variax Hexacenter (Figure from Giddings and Lewis)

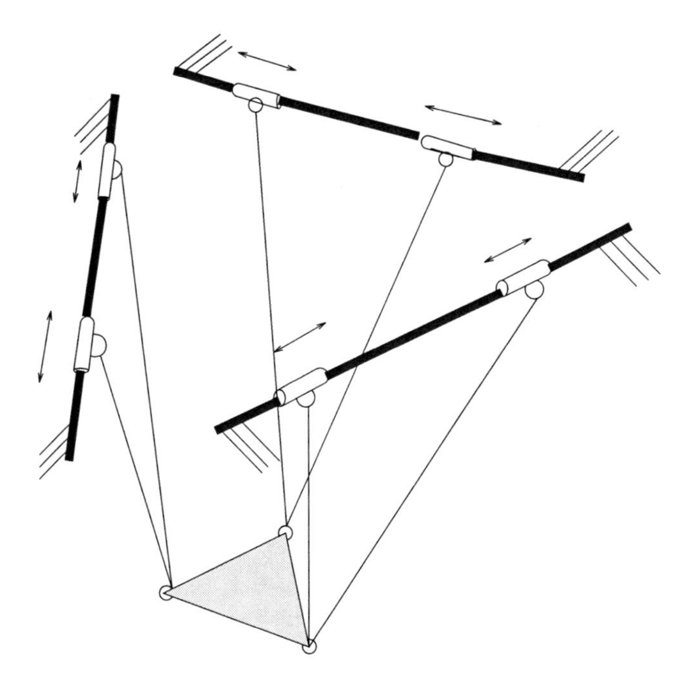

Fig. 1.14 Kinematic structure of the 6-dof machine tools

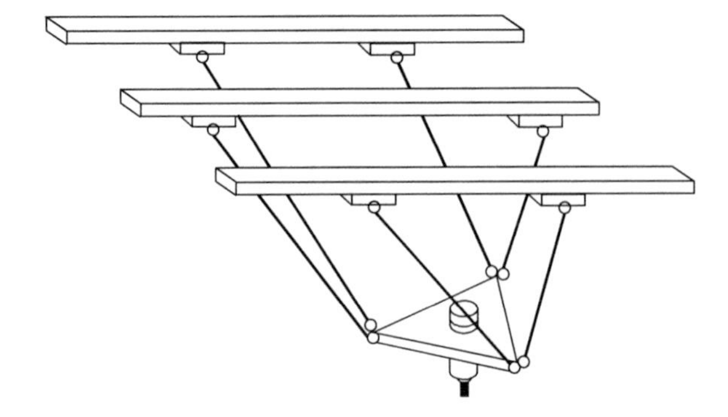

Fig. 1.15 Kinematic structure of the Hexaglide

- Linear Motion (LM) actuated machines with fixed leg lengths and base joints movable on linear guideways (e.g., HexaM, ECLIPSE, Hexaglide, Triaglide, Linapod),
- Revolute actuated machines with fixed leg lengths (e.g., Delta, Hexa),

5. From the viewpoint of research methodology, there are OKP (One-of-a-Kind Production) design methodology (e.g., Tricept, HexaM), which is suitable for those industrial companies, and systematic design of Hexapods using modular robot methodology (e.g., Linapod). Modular robot concepts and techniques have been of interest in the robotics field since the 1980s [32, 37], since selecting an industrial robot that will best suit the needs of a forecast set of tasks can be a difficult and costly exercise. This problem can be alleviated by using a modular robot (system) that consists of standard units such as joints and links, which can be efficiently configured into the most suitable leg geometry for these tasks. From this point of view, modular robots introduce a new dimension to flexible automation in terms of hardware flexibility, compared with conventional industrial robots.

Figure 1.16 shows some of the possible configurations of parallel kinematic mechanisms that can be found primarily in [1]. The patented machine tools in Fig. 1.16a "Hexa" [147] and Fig. 1.16b "Rotary Hexapod" [34] are revolute actuated ones while Fig. 1.16c 6-dof parallel mechanism [9] and Fig. 1.16d "Eclipse" [132] are the combination of revolute and prismatic actuated mechanisms. Figure 1.16e 6-dof "minimanipulator" [140] uses 2 prismatic actuators with fixed leg lengths and Fig. 1.16f [18] displays the combination of a linear driven base point and variable strut lengths.

Philosophically, most of the work above was built upon the concept of the traditional "Gough-Stewart" mechanism type. This suggests that most parallel mechanisms have six degrees of freedom. A question left open in previous work is: The vast majority of the machining is done with less than 6-dof, so why should we pay for six? In this book, we will focus our attention on 5-dof or less than 5-dof parallel mechanisms (Fig. 1.17), since machining consists in orienting an axisymmetric body (the tool), which requires only five degrees of freedom.

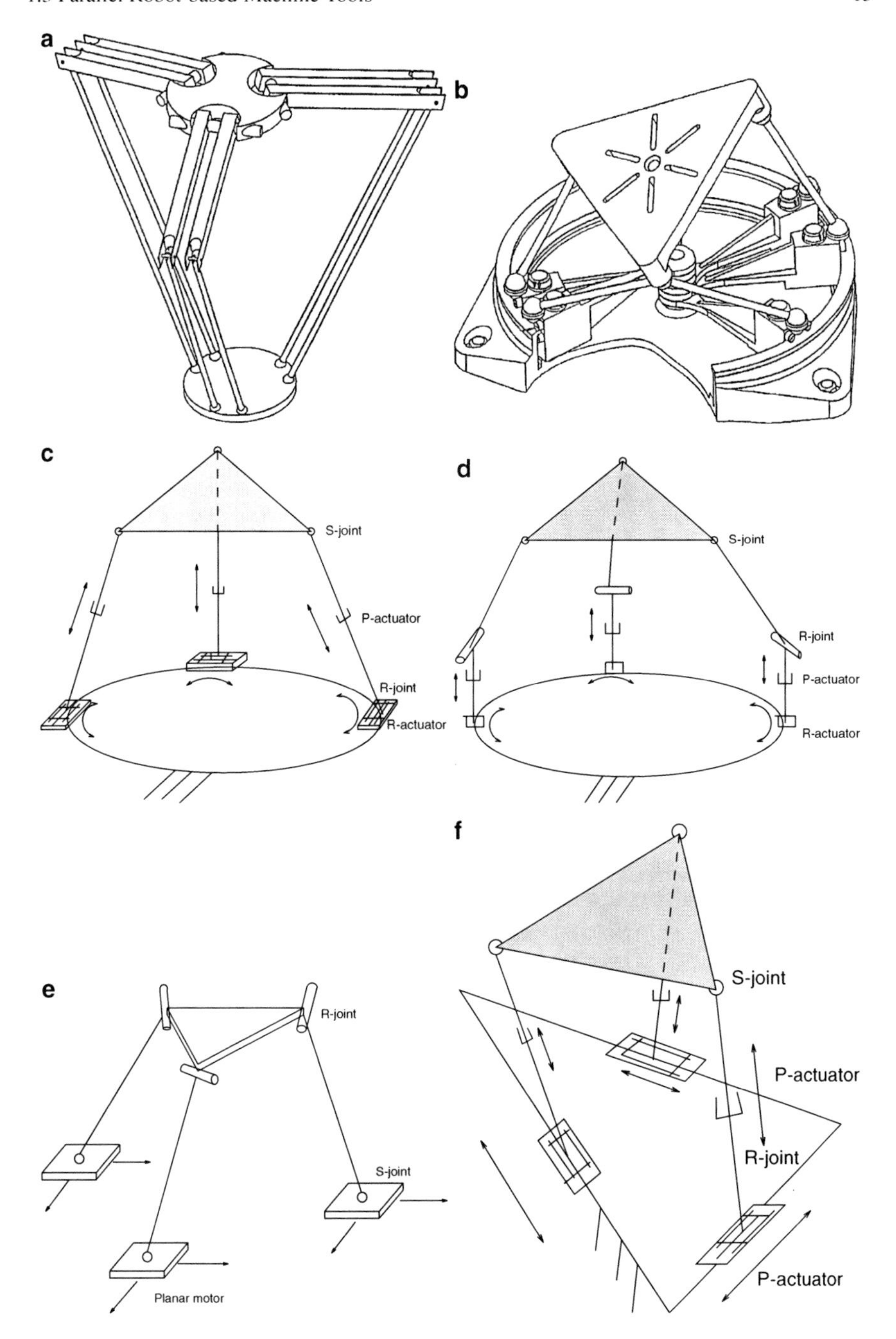

Fig. 1.16 Selected parallel kinematic mechanisms. (**a**) The "Hexa" robot (Uchiyama 1994). (**b**) The "Rotary Hexapod" by Hexel (Chi 1999). (**c**) Circular movement of the base point. (**d**) "Eclipse" from SNU. (**e**) 6-dof "minimanipulator". (**f**) Combination of linear driven base point and variable strut length

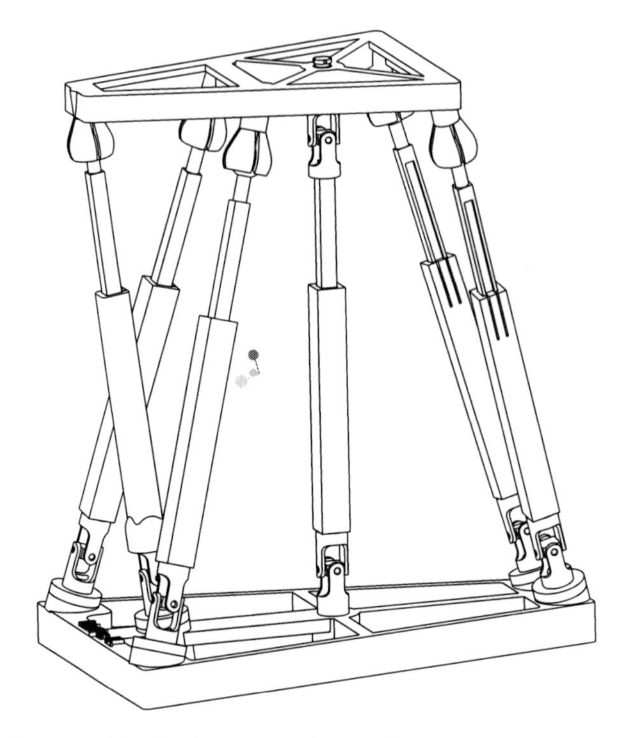

Fig. 1.17 CAD model of the 5-dof parallel mechanism (Figure by Gabriel Coté)

In this book, we propose a series of n-dof parallel mechanisms which consist of n identical actuated legs with six degrees of freedom and one passive leg with n degrees of freedom connecting the platform and the base. The degree of freedom of the mechanism is dependent on the passive leg's degree of freedom. One can improve the rigidity of this type of mechanism through optimization of the link rigidities to reach the maximal global stiffness and precision.

1.4 Scope and Organization of this Book

Conventional machine tools are usually based on a serial structure. There are as many degrees of the freedom as required, and the axes are arranged in series. This leads to a single kinematic chain. The axes are usually arranged according to the Cartesian axes, which means there is a X, Y, and Z axis and rotational axes if needed. These machines are easy to operate because each axis directly controls one Cartesian degree of freedom and there is no coupling between the axes.

A parallel kinematic machine promises to increase stiffness, higher speed, and acceleration due to reduced moving mass, reduced production, and installation costs. Research in this kind of architectures for machine tools has been growing since the 1980s [12,15,70]. Although a number of new devices were patented, none seems to take the structure flexibility into account. Although the joints and links have become commercially available, the study for the most promising architecture for machine tools through kinetostatic analysis, dynamics, and optimization is still a challenge. The aim of this book is to provide readers the new alternative mechanical architectures which could be used in the design of a machine tool with parallel or hybrid architecture. To reach this goal, the objectives are set as follows:

1. Development of a topological representation and generation of all possible architectures that will provide 5 degrees of freedom between the tool and the workpiece. The topological representation serves to develop a database for conceptual design to obtain the most promising kinematic architectures for 5-dof or fewer than 5-dof machine tools. The key consideration in achieving this objective are (1) both the tool and the workpiece can be actuated independently and 5 dofs are required for manufacturing tasks, (2) the possible combinations of 5 dofs are: (5, 0), (4, 1), and (3, 2), and (3) for each of these combinations, the kinematic chains involved may lead to several possibilities (serial, parallel, or hybrid) and additionally, redundancy may be an option. At the end of this study, a detailed list of possible topologies will be obtained and the most promising architectures will be highlighted.
2. Development of geometric design model. The key task is for the topologies selected in the previous study, to define geometric parameters and investigate the geometric design. The geometric design must take into account the actuation issues, the working volume, and mechanical interferences. The selected designs will be modeled using Pro-Engineer, which will facilitate this step. Again, all possibilities of configurations will be investigated.
3. Development of a general model of the stiffness of the mechanisms screened out from the list of some promising configurations. Using a formulation based on lumped flexibilities, write a general model of the stiffness of the concerned mechanisms. Using this model, all concerned mechanisms will be analyzed for their stiffness and accuracy at the tool, which is the most important property of the mechanism. In the lumped model, links and actuators will be replaced by springs whose stiffness will represent the stiffness of the link or the actuator.
4. Development of the kinetostatic modeling for parallel robot design. Kinematic/static duality can be derived by considering the power input to and output from a system, which can neither store nor dissipate energy, namely, a system in which kinetic energy, strain energy, friction and damping are all absent and where gravitational forces are considered as external forces applied to the system. Thus, term "Kinetostatic Analysis" as such: Given the mechanism motion, calculate the unknown internal joint forces and external input forces or torques. Kinetostatic analysis includes two analyzes: (1) kinematic solutions to provide the mechanism motion, (2) stiffness solutions to relate the forces and torques to the motion. Using the kinetostatic model developed in the preceding step, the most promising architectures for stiffness (accuracy) based on constraints associated with size and geometry can be optimized.

5. Development of reconfigurable parallel robotic machine tools. The new design uses an adjustable architecture, so the machine tool has the capability to machine all five sides of a workpiece, and adjust the depth, clearance of interference and dynamic performance. The proposed reconfigurable 5-DOF parallel kinematic machine can machine 5 sides of the workpiece freely with a simple mechanism, hence providing savings in the motion system construction and implementation. The reconfigurable system was implemented by an adjusting architecture. The findings have significant potential for industrial applications.
6. Development of the synthetical methodology for performance evaluation and design optimization. The mean value and the standard deviation of the stiffness distribution are proposed as the design indices. The mean value represents the average stiffness of the parallel robot manipulator over the workspace, while the standard deviation indicates the stiffness variation relative to the mean value. In general, the higher the mean value the less the deformation and the lower the standard deviation, the more uniform the stiffness distribution over the workspace. A design optimization based on these global stiffness indices is further investigated. At this point, a multi-objective optimization issue will be defined. Genetic algorithms based Pareto optimal frontier set in the solution space can be obtained as the results of comprehensive stiffness design, and other performance indices are also considered.
7. Development of the optimal calibration method. It is known that calibration is best performed in the least sensitive error region within an entire workspace. Because of the complexity of the error sources, it is difficult to develop the calibration model if all the errors will be considered. Errors including manufacturing and assembly error, thermal error, and nonlinear stiffness error are considered as a single error source (pseudo-error source), which only causes the deviation of joint variables. Artificial neural network will be applied to describe the complex nonlinear relationship between joint variables and deviation of joint variables with respect to the measured pose of the end-effector. The pseudo-error in arbitrary joint variable can be obtained and thus the control parameters can be adjusted accordingly.
8. Development integrated environment of parallel manipulator-based machine. The system included a kinematic/dynamic analysis, kinetostatic modeling, CAD module, FEM module, CAM module, optimization module and a visual environment for simulation and collision detection of the machining and deburring process. It represents an integration for the design, analysis, optimization, and simulation of the parallel kinematic machine. An approach for web-based real-time monitoring and remote control is also developed. The effectiveness of the system is shown through results obtained by the National Research Council of Canada, during the design of a 3-DOF Tripod parallel robotic machine. Notable advantages of the new system included ease and efficiency, thereby allowing a real-time simulation.

Although the proposed investigation is aimed at the most promising 5-dof or fewer than 5-dof machine tools architectures, those issues addressed in the eight objectives are fundamental. Therefore, the results of the work can provide a framework for facilitating a further study of parallel mechanisms for machine tools.

Chapter 2
Kinematics of Mechanisms

2.1 Preamble

Robot kinematics is the study of the motion (kinematics) of robotic mechanisms. In a kinematic analysis, the position, velocity, and acceleration of all the links are calculated with respect to a fixed reference coordinate system, without considering the forces or moments. The relationship between motion and the associated forces and torques is studied in robot dynamics. Forward kinematics and inverse kinematics are the main components in robot kinematics.

Forward kinematics (also known as direct kinematics) is the computation of the position and orientation of a robot's end effector as a function of its joint angles. Inverse kinematics is defined as: given the position and orientation of a robot's end-effector, calculate all possible sets of joint motion that could be used to attain this given position and orientation.

From the viewpoint of robot structure, robot can be divided into two basic types: serial robot and parallel robot. Besides, there is a hybrid type, which is the combination of serial and parallel robots. Serial robots have open kinematic chain, which can be further classified as either articulated or cartesian robots.

In the following, the basic mathematical and geometric concepts including position and orientation of a rigid body are presented (Sect. 2.2). Translational coordinate transformation, rotational coordinate transformation and homogeneous transformation are introduced in Sect. 2.3. Denavit–Hartenberg expression of kinematic parameters is discussed in Sect. 2.4. Section 2.5 describes the derivation of Jacobian Matrix. Finally, the conclusions are given in Sect. 2.6.

2.2 Position and Orientation of Rigid Body

2.2.1 Rotation Matrix

To explain the relationship between parts, tools, manipulator etc., some concepts such as position vector, plane, and coordinate frame are utilized.

D. Zhang, *Parallel Robotic Machine Tools*, DOI 10.1007/978-1-4419-1117-9_2,

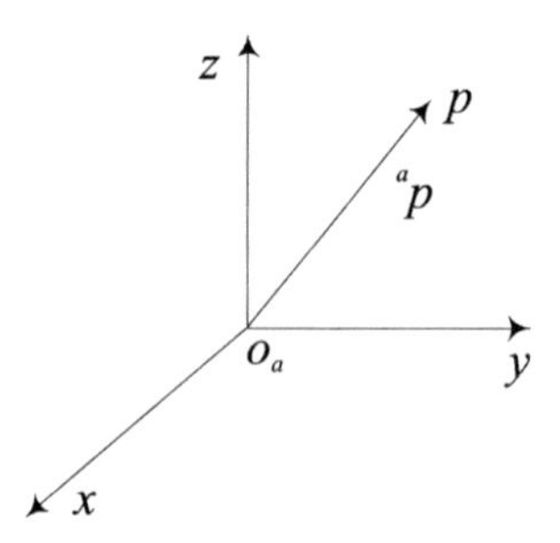

Fig. 2.1 Presentation of position

The motion of a robot can be described by its position and orientation, which is called pose as well. Once the reference coordinate system has been established, any point in the space can be expressed by a (3×1) vector. For orthogonal coordinate system $\{O_a - x_a y_a z_a\}$, any point $\boldsymbol{p}$ in the space can be written as follow:

$$ {}^a\boldsymbol{p} = \begin{bmatrix} p_x \\ p_y \\ p_z \end{bmatrix}, \tag{2.1} $$

where p_x, p_y, p_z denote the components of the vector $\boldsymbol{p}$ along the coordinate axis x_a, y_a, z_a, respectively. Here, $\boldsymbol{p}$ is called position vector, which is shown in Fig. 2.1.

To investigate the motion and manipulation of robots, not only the description of position is needed, but also the orientation is likewise important. To define the orientation of point b, we should assume that there is an orthogonal coordinate system $\{O_b - x_b y_b z_b\}$ attached to the point. Here, x_b, y_b, z_b denote the unit vectors of the coordinate axes. With respect to the reference coordinate system $\{O_a - x_a y_a z_a\}$, the orientation of point b is expressed as follow:

$$ {}^a_b\boldsymbol{R} = \left[{}^a\mathbf{x}_b \; {}^a\mathbf{y}_b \; {}^a\mathbf{z}_b\right] = \begin{bmatrix} r_{11} & r_{12} & r_{13} \\ r_{21} & r_{22} & r_{23} \\ r_{31} & r_{32} & r_{33} \end{bmatrix}, \tag{2.2} $$

where ${}^a_b\boldsymbol{R}$ is called rotation matrix. ${}^a_b\boldsymbol{R}$ has nine elements in total, but only three of them are independent. The following constraint conditions should be satisfied by the nine elements:

$$ {}^a\mathbf{x}_b \cdot {}^a\mathbf{x}_b = {}^a\mathbf{y}_b \cdot {}^a\mathbf{y}_b = {}^a\mathbf{z}_b \cdot {}^a\mathbf{z}_b = 1, \tag{2.3} $$

$$ {}^a\mathbf{x}_b \cdot {}^a\mathbf{y}_b = {}^a\mathbf{y}_b \cdot {}^a\mathbf{z}_b = {}^a\mathbf{z}_b \cdot {}^a\mathbf{x}_b = 0. \tag{2.4} $$

It can be concluded that the rotation matrix ${}^a_b\boldsymbol{R}$ is orthogonal, and the following condition should be satisfied:

$$ {}^a_b\boldsymbol{R}^{-1} = {}^a_b\boldsymbol{R}^T; \quad |{}^a_b\boldsymbol{R}| = 1. \tag{2.5} $$

The rotation matrix with respect to the rotation transformation by an angle θ about the axis x, y, z, respectively, can be calculated:

$$\boldsymbol{R}(x,\theta) = \begin{bmatrix} 1 & 0 & 0 \\ 0 & c\theta & -s\theta \\ 0 & s\theta & c\theta \end{bmatrix}, \tag{2.6}$$

$$\boldsymbol{R}(y,\theta) = \begin{bmatrix} c\theta & 0 & s\theta \\ 0 & 1 & 0 \\ -s\theta & 0 & c\theta \end{bmatrix}, \tag{2.7}$$

$$\boldsymbol{R}(z,\theta) = \begin{bmatrix} c\theta & -s\theta & 0 \\ s\theta & c\theta & 0 \\ 0 & 0 & 1 \end{bmatrix}, \tag{2.8}$$

where $s\theta = \sin\theta$ and $c\theta = \cos\theta$

Suppose that coordinate frames $\{\mathbf{B}\}$ and $\{\mathbf{A}\}$ have the same orientation. But the original points of the two coordinate frames do not overlap. Using the position vector ${}^a\boldsymbol{p}_{O_b}$ to describe the position related to frame $\{\mathbf{A}\}$. ${}^a\boldsymbol{p}_{O_b}$ is called the translational vector of frame $\{\mathbf{B}\}$ with respect to frame $\{\mathbf{A}\}$. If the position of point $\boldsymbol{p}$ in the coordinate frame $\{\mathbf{B}\}$ is written as ${}^b\boldsymbol{p}$, then the position vector of $\boldsymbol{p}$ with respect to frame $\{\mathbf{A}\}$ can be written as follows:

$${}^a\boldsymbol{p} = {}^b\boldsymbol{p} + {}^a\boldsymbol{p}_{O_b}, \tag{2.9}$$

That is equation of coordinate translation which is shown in Fig. 2.2.

Suppose that coordinate frames $\{\mathbf{B}\}$ and $\{\mathbf{A}\}$ have the same orientation, but their orientation is different. Using the rotation matrix ${}^a_b\boldsymbol{R}$ to describe the orientation of

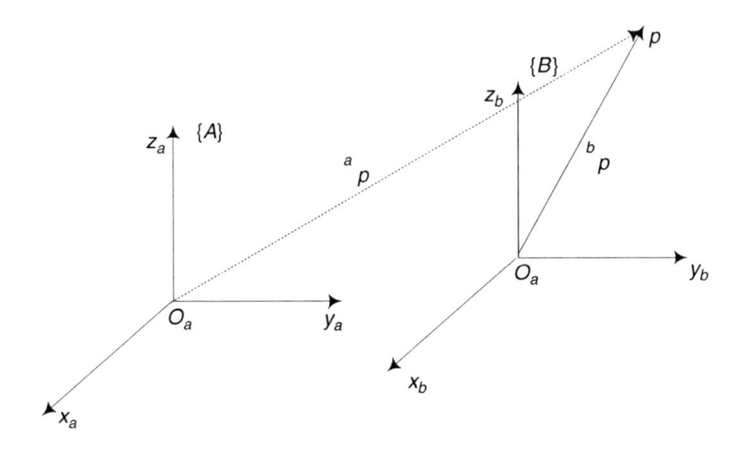

Fig. 2.2 Translational transformation

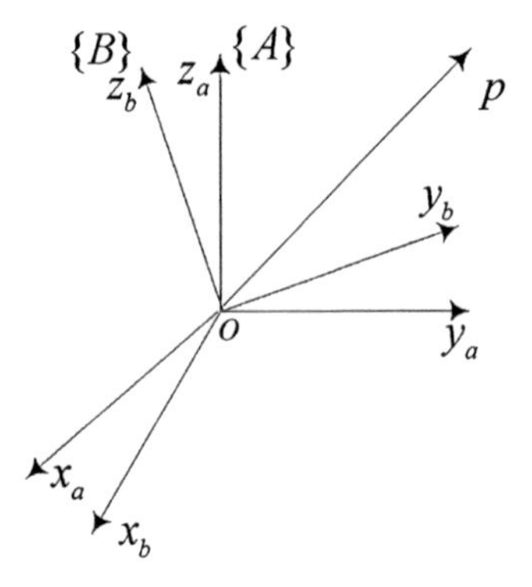

Fig. 2.3 Rotational transformation

frame {**B**} with respect to frame {**A**}, then the transformation of point $\boldsymbol{p}$ in frames {**A**} and {**B**} can be deduced as:

$$ {}^{a}\boldsymbol{p} = {}^{a}_{b}\boldsymbol{R} \cdot {}^{b}\boldsymbol{p}, \tag{2.10} $$

where ${}^{a}\boldsymbol{p}$ denotes the position $\boldsymbol{p}$ with the reference coordinate system {**A**}, and ${}^{b}\boldsymbol{p}$ denotes the position $\boldsymbol{p}$ with the reference coordinate system {**B**}. It is called equation of coordinate rotation which is shown in Fig. 2.3.

The following equation can be deduced:

$$ {}^{b}_{a}\boldsymbol{R} = {}^{a}_{b}\boldsymbol{R}^{-1} = {}^{a}_{b}\mathbf{R}^{-T}. \tag{2.11} $$

For the common condition, neither the original points of frames {**A**} and {**B**} overlap nor they have the same orientation. Use the position vector ${}^{a}\boldsymbol{p}_{O_b}$ to describe the original point of frame {**B**} with respect to frame {**A**}; use the rotation matrix ${}^{a}_{b}\boldsymbol{R}$ to describe the orientation of frame {**B**} with respect to frame {**A**}. To any point in the space, the transformation can be found:

$$ {}^{a}\boldsymbol{p} = {}^{a}_{b}\boldsymbol{R} \cdot {}^{b}\boldsymbol{p} + {}^{a}\boldsymbol{p}_{O_b}. \tag{2.12} $$

2.2.2 Euler Angles

The Euler angle I, shown in Fig. 2.4, defines a rotation angle ϕ around the z-axis, then a rotation angle θ around the new x-axis, and a rotation angle φ around the new z-axis.

$$ \boldsymbol{R}_{z\phi} = \begin{bmatrix} c\phi & -s\phi & 0 \\ s\phi & c\phi & 0 \\ 0 & 0 & 1 \end{bmatrix}, \quad \boldsymbol{R}_{u'\theta} = \begin{bmatrix} 1 & 0 & 0 \\ 0 & c\theta & -s\theta \\ 0 & s\theta & c\theta \end{bmatrix}, \quad \boldsymbol{R}_{w''\varphi} = \begin{bmatrix} c\varphi & -s\varphi & 0 \\ s\varphi & c\varphi & 0 \\ 0 & 0 & 1 \end{bmatrix}. \tag{2.13} $$

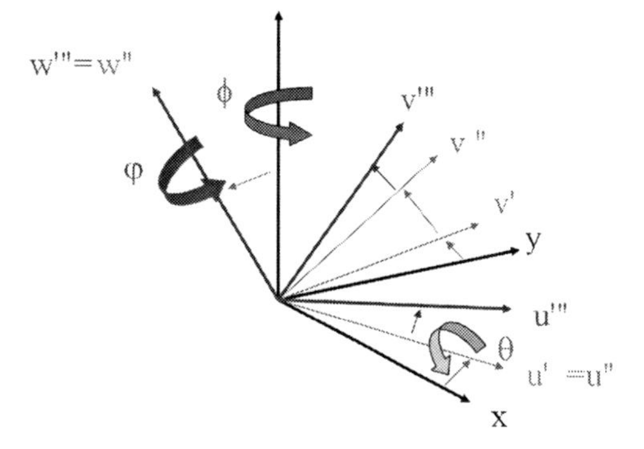

Fig. 2.4 Euler angle I

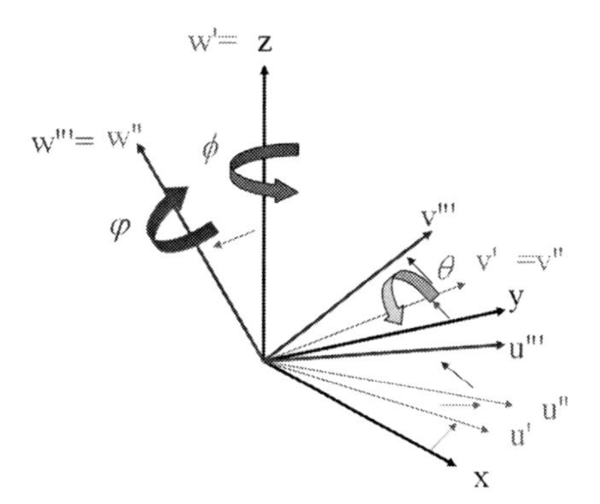

Fig. 2.5 Euler angle II

Resultant Eulerian rotation matrix generates:

$$\boldsymbol{R} = \boldsymbol{R}_{z\phi}\boldsymbol{R}_{u'\theta}\boldsymbol{R}_{w''\varphi} = \begin{bmatrix} c\phi c\varphi - s\phi s\varphi c\theta & -c\phi s\varphi - s\phi c\varphi c\theta & s\varphi s\theta \\ s\phi c\varphi + c\phi s\varphi c\theta & -s\phi s\varphi + c\phi c\varphi c\theta & -c\phi s\theta \\ s\varphi s\theta & c\varphi s\theta & c\theta \end{bmatrix}. \tag{2.14}$$

The Euler angle II, shown in Fig. 2.5, defines a rotation of angle ϕ around the z-axis, then a rotation of angle θ around the new y-axis, and finally a rotation angle φ around the new z-axis.

Note the opposite (clockwise) sense of the third rotation ϕ. Matrix with Euler Angle II generates:

$$\begin{bmatrix} -s\phi s\varphi + c\phi c\varphi c\theta & -s\phi c\varphi - s\phi c\varphi c\theta & c\phi s\theta \\ c\phi s\varphi + s\phi c\varphi c\theta & c\phi c\varphi - s\phi c\varphi c\theta & s\varphi s\theta \\ -c\varphi s\theta & s\varphi s\theta & c\theta \end{bmatrix} \tag{2.15}$$

2.3 Homogeneous Transformation

If the coordinates of any point in an orthogonal coordinate system is given, then the coordinates of this point in another orthogonal coordinate system can be calculated by homogeneous coordinate transformation.

The transformation (2.12) is inhomogeneous to point ${}^{b}\boldsymbol{p}$, but it can be expressed by an equivalent homogeneous transformation:

$$\begin{bmatrix} {}^{a}\boldsymbol{p} \\ 1 \end{bmatrix} = \begin{bmatrix} {}^{a}_{b}\boldsymbol{R} & {}^{a}\boldsymbol{p}_{O_b} \\ 0_{1\times 3} & 1 \end{bmatrix} = \begin{bmatrix} {}^{b}\boldsymbol{p} \\ 1 \end{bmatrix}, \tag{2.16}$$

where the vector (4×1) denotes the coordinates in three-dimensional space. It still can be noted as ${}^{a}\boldsymbol{p}$ or ${}^{b}\boldsymbol{p}$. The above equation can be rewritten in the format of matrix:

$${}^{a}\boldsymbol{p} = {}^{a}_{b}\boldsymbol{T} \cdot {}^{b}\boldsymbol{p} + {}^{a}\boldsymbol{p}_{O_b}, \tag{2.17}$$

where the vector (4×1) of ${}^{a}\boldsymbol{p}$ and ${}^{b}\boldsymbol{p}$ is called homogeneous coordinates, here,

$${}^{a}_{b}\boldsymbol{T} = \begin{bmatrix} {}^{a}_{b}\boldsymbol{R} & {}^{a}\boldsymbol{p}_{O_b} \\ 0_{4\times 1} & 1 \end{bmatrix}. \tag{2.18}$$

In fact, the transformation (2.18) is equivalent to (2.12). The (2.17) can be rewritten as

$${}^{a}\boldsymbol{p} = {}^{a}_{b}\boldsymbol{R} \cdot {}^{b}\boldsymbol{p} + {}^{a}\boldsymbol{p}_{O_b}. \tag{2.19}$$

Suppose vector $ai + bj + ck$ describes one point in the space, where i, j, k are the unit vector of the axes x, y, z, respectively. This point can be expressed by the translational homogeneous transformation matrix.

$$Trans(a,b,c) = \begin{bmatrix} 1 & 0 & 0 & a \\ 0 & 1 & 0 & b \\ 0 & 0 & 0 & c \\ 0 & 0 & 0 & 1 \end{bmatrix}, \tag{2.20}$$

where Trans denotes translational transformation.

If a rigid body rotates about x, y and z-axis with θ, then the following equations can be obtained:

$$Rot(x,\theta) = \begin{bmatrix} 1 & 0 & 0 & 0 \\ 0 & c\theta & -s\theta & 0 \\ 0 & s\theta & c\theta & 0 \\ 0 & 0 & 0 & 1 \end{bmatrix}, \tag{2.21}$$

$$Rot(y,\theta) = \begin{bmatrix} c\theta & 0 & s\theta & 0 \\ 0 & 1 & 0 & 0 \\ -s\theta & 0 & c\theta & 0 \\ 0 & 0 & 0 & 1 \end{bmatrix}, \tag{2.22}$$

$$Rot(z,\theta) = \begin{bmatrix} c\theta & -s\theta & 0 & 0 \\ s\theta & c\theta & 0 & 0 \\ 0 & 0 & 1 & 0 \\ 0 & 0 & 0 & 1 \end{bmatrix}, \tag{2.23}$$

where *Rot* denotes rotational transformation.

As the transformation is based on the fixed reference frame, a left-handed multiplication of transformation sequences is followed. For example, a rigid body rotates 90° about the z-axis of the reference frame, then it rotates another 90° about the y-axis and finally it translates 4 unit lengths along x-axis of the fixed reference frame, the transformation of this rigid body can be described as:

$$\boldsymbol{T} = Trans(4,0,0)Trans(y,90)Trans(z,90) = \begin{bmatrix} 0 & 0 & 1 & 4 \\ 1 & 0 & 0 & 0 \\ 0 & 1 & 0 & 0 \\ 0 & 0 & 0 & 1 \end{bmatrix}. \tag{2.24}$$

The above matrix represents the operations of rotation and translation about the primary reference frame. The six points of the wedge-shaped object (Fig. 2.6(a)) can be expressed as:

$$\begin{bmatrix} 0 & 0 & 1 & 4 \\ 1 & 0 & 0 & 0 \\ 0 & 1 & 0 & 0 \\ 0 & 0 & 0 & 1 \end{bmatrix} \begin{bmatrix} 1 & -1 & -1 & 1 & 1 & -1 \\ 1 & 0 & 0 & 0 & 4 & 4 \\ 0 & 0 & 2 & 2 & 0 & 0 \\ 1 & 1 & 1 & 1 & 1 & 1 \end{bmatrix} = \begin{bmatrix} 4 & 4 & 6 & 6 & 4 & 4 \\ 4 & -1 & -1 & 1 & 1 & -1 \\ 0 & 0 & 2 & 2 & 4 & 4 \\ 1 & 1 & 1 & 1 & 1 & 1 \end{bmatrix}. \tag{2.25}$$

Figure 2.6(b) shows the result of transformation.

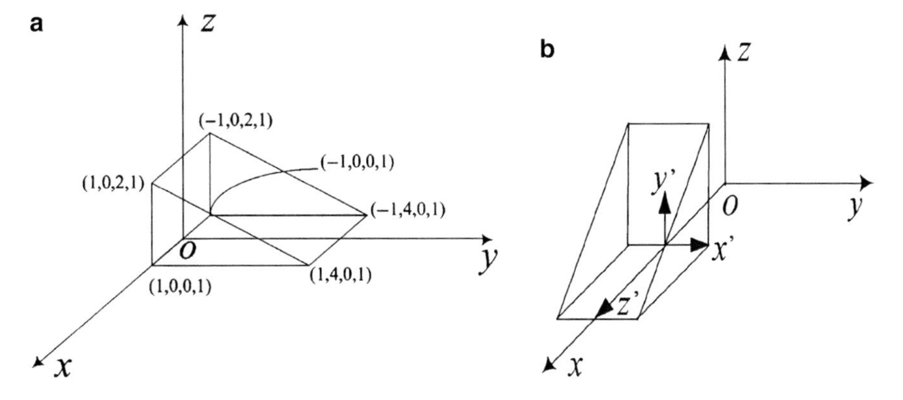

Fig. 2.6 Transformation of wedge-shaped object

In the above sections, the rotational transformation matrix with respect to rotations about x, y and z-axis has been analyzed. Here is the rotation matrix in the common situation: rotation about any vector (axis) with θ.

Suppose **f** is the unit vector of z-axis in coordinate frame C, namely:

$$\mathbf{C} = \begin{bmatrix} n_x & o_x & a_x & 0 \\ n_y & o_y & a_y & 0 \\ n_z & o_z & a_z & 0 \\ 0 & 0 & 0 & 1 \end{bmatrix}, \tag{2.26}$$

$$\mathbf{f} = a_x i + a_y j + a_z k. \tag{2.27}$$

Therefore, rotation about vector f is equivalent to rotation about z-axis in coordinate frame C, thus one has,

$$Rot(\mathbf{f}, \theta) = Rot(\mathbf{c}, \theta). \tag{2.28}$$

If the coordinate frame $\{T\}$ is known with respect to reference coordinate frame, then another coordinate frame $\{S\}$ can be calculated with respect to frame $\{C\}$, because,

$$T = CS, \tag{2.29}$$

Where, S is the relative position of T with respect to C, then,

$$S = C^{-1}T. \tag{2.30}$$

The rotation of T about f is equivalent to the rotation of S about z-axis of frame $\{C\}$,

$$Rot(\mathbf{f}, \theta)T = CRot(\mathbf{z}, \theta)S, \tag{2.31}$$

$$Rot(\mathbf{f}, \theta)T = CRot(\mathbf{z}, \theta)C^{-1}T. \tag{2.32}$$

Then the following equation can be derived,

$$Rot(\mathbf{f}, \theta) = CRot(\mathbf{z}, \theta)C^{-1}. \tag{2.33}$$

As **f** is the z-axis of frame $\{C\}$, then it can be found that $Rot(\mathbf{z}, \theta)C^{-1}$ is just the function of **f**, because,

$$CRot(\mathbf{z}, \theta)C^{-1} = \begin{bmatrix} n_x & o_x & a_x & 0 \\ n_y & o_y & a_y & 0 \\ n_z & o_z & a_z & 0 \\ 0 & 0 & 0 & 1 \end{bmatrix} \begin{bmatrix} c\theta & -s\theta & 0 & 0 \\ s\theta & c\theta & 0 & 0 \\ 0 & 0 & 1 & 0 \\ 0 & 0 & 0 & 1 \end{bmatrix} \begin{bmatrix} n_x & o_x & a_x & 0 \\ n_y & o_y & a_y & 0 \\ n_z & o_z & a_z & 0 \\ 0 & 0 & 0 & 1 \end{bmatrix}^{-1}. \tag{2.34}$$

Note that $z = a$, $vers\theta = 1 - c\theta$, $f = z$. Equation (2.34) can be simplified as,

$$Rot(\mathbf{f},\ \theta) = \begin{bmatrix} \mathbf{f}_x\mathbf{f}_x vers\theta + c\theta & \mathbf{f}_y\mathbf{f}_x vers\theta - \mathbf{f}_z s\theta & \mathbf{f}_z\mathbf{f}_x vers\theta + \mathbf{f}_y s\theta & 0 \\ \mathbf{f}_x\mathbf{f}_y vers\theta + \mathbf{f}_z s\theta & \mathbf{f}_y\mathbf{f}_y vers\theta + c\theta & \mathbf{f}_z\mathbf{f}_y vers\theta - \mathbf{f}_x s\theta & 0 \\ \mathbf{f}_x\mathbf{f}_z vers\theta + \mathbf{f}_z s\theta & \mathbf{f}_y\mathbf{f}_z vers\theta + \mathbf{f}_x s\theta & \mathbf{f}_z\mathbf{f}_z vers\theta + c\theta & 0 \\ 0 & 0 & 0 & 1 \end{bmatrix}. \tag{2.35}$$

Each basic rotation transformation can be derived from the general rotation transformation, i.e., if $\mathbf{f}_x = 1, \mathbf{f}_y = 0$ and $\mathbf{f}_z = 0$, then $Rot(\mathbf{f},\ \theta) = Rot(\mathbf{x},\ \theta)$. Equation (2.35) yields,

$$Rot(x, \theta) = \begin{bmatrix} 1 & 0 & 0 & 0 \\ 0 & c\theta & -s\theta & 0 \\ 0 & s\theta & c\theta & 0 \\ 0 & 0 & 0 & 1 \end{bmatrix}, \tag{2.36}$$

which is identical to (2.21).

2.4 Denavit–Hartenberg Representation

Denavit–Hartenberg (DH) representation is a generic and simple method to define the relative motion parameters of two consecutive links and joints. Any arbitrary type of mechanism can be represented using the DH method to relate the position and orientation of the last link to the first one.

In studying the kinematic motion between two jointed links, the DH method defines the position and orientation of two consecutive links in a chain, link i with respect to link $(i - 1)$ using a 4×4 homogeneous transformation matrix. With reference to Fig. 2.7, let axis i denotes the axis of the joint connecting link $(i - 1)$ to link i. Four parameters should be determined, which are θ_i, d_i, a_i, and α_i. θ_i denotes the rotation angle measured from axis x_{i-1} to x_i with respect to z_i axis. d_n denotes the displacement measured from axis x_{n-1} to x_n with respect to z_n axis. a_i denotes the displacement measured from axis z_i to z_{i+1} with respect to x_i axis. α_i denotes the rotation angle measured from axis z_i to z_{i+1} with respect to x_i axis.

Following steps can help to determine the link and joint parameters of the whole kinematic model.

- Number the joints from 1 to n starting with the base and ending with the end-effecter
- Establish the base coordinate system. Establish a right-handed orthonormal coordinate system (x_0, y_0, z_0) at the supporting base with z_0 axis lying along the axis of motion of joint 1
- Establish joint axis. Align the z_i with the axis of motion (rotary or sliding) of joint $i + 1$

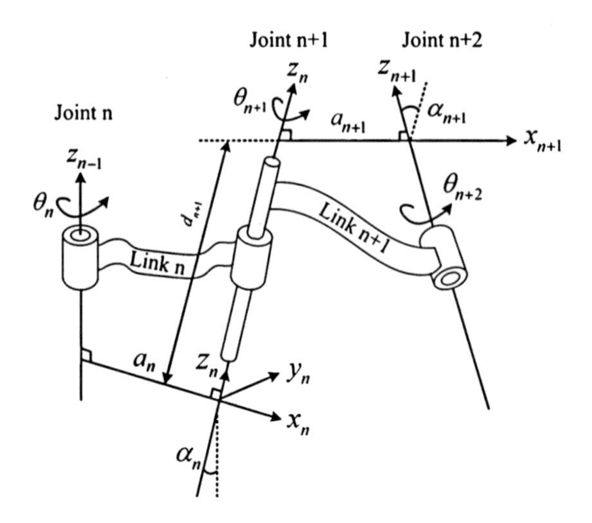

Fig. 2.7 Denavit–Hartenberg kinematic description

- Establish the origin of the ith coordinate system. Locate the origin of the ith coordinate at the intersection of the z_i and z_{i-1} or at the intersection of common normal between the z_i and z_{i-1} axes and the z_i axis
- Establish x_i axis. Establish $x_i = \pm(z_{i-1} \times z_i)/\|z_{i-1} \times z_i\|$ or along the common normal between the z_{i-1} and z_i axes when they are parallel
- Establish y_i axis. Assign $y_i = +(z_i \times x_i)/\|z_i \times x_i\|$ to complete the right-handed coordinate system
- Find the link and joint parameters

With the parameters defined in Fig. 2.5, the DH model transformation matrix can be obtained as follows

$$ {}^{i-1}_{i}\mathbf{T} = \mathbf{A}_i = Rot(z,\ \theta_i) \times Trans(0,0,d_i) \times Trans(a_i,0,0) \times Rot(x,\alpha_i), \quad (2.37) $$

$$ \mathbf{A}_i = \begin{bmatrix} c\theta_i & -s\theta_i c\alpha_i & s\theta_i s\alpha_i & a_i c\theta_i \\ s\theta_i & c\theta_i c\alpha_i & -c\theta_i s\alpha_i & a_i s\theta_i \\ 0 & s\alpha_i & c\alpha_i & d_i \\ 0 & 0 & 0 & 1 \end{bmatrix}. \quad (2.38) $$

2.5 Jacobian Matrix

To describe a micro-motion of robot, differential coefficient is utilized for coordinate transformation. Given a coordinate frame {T},

$$ T + dT = Trans(dx, dy, dz) Rot(f, d\theta) T, \quad (2.39) $$

where $Trans(dx, dy, dz)$ denotes the differential translation of dx, dy, dz, and $Rot(f, d\theta)$ denotes the differential rotation about the vector f. Then dT can be calculated as follows:

$$dT = [Trans(dx, dy, dz)Rot(f, d\theta) - I]T. \tag{2.40}$$

The homogeneous transformation expressing differential translation is

$$Trans(dx, dy, dz) = \begin{bmatrix} 1 & 0 & 0 & dx \\ 0 & 1 & 0 & dy \\ 0 & 0 & 1 & dz \\ 0 & 0 & 0 & 1 \end{bmatrix}. \tag{2.41}$$

For the formula of general rotation transformation

$$Rot(f, d\theta) = \begin{bmatrix} f_x f_x vers\theta + c\theta & f_y f_x vers\theta - f_z s\theta & f_z f_x vers\theta + f_y s\theta & 0 \\ f_x f_y vers\theta + f_z s\theta & f_y f_y vers\theta + c\theta & f_z f_y vers\theta - f_x s\theta & 0 \\ f_x f_z vers\theta + f_z s\theta & f_y f_z vers\theta + f_x s\theta & f_z f_z vers\theta + c\theta & 0 \\ 0 & 0 & 0 & 1 \end{bmatrix}. \tag{2.42}$$

Since $\lim_{\theta \to 0} \sin\theta = d\theta$, $\lim_{\theta \to 0} \cos\theta = 1$, $\lim_{\theta \to 0} vers\ \theta = 0$, differential rotational homogeneous transformation can be expressed as,

$$Rot(f, d\theta) = \begin{bmatrix} 1 & -f_z d\theta & f_y d\theta & 0 \\ f_z d\theta & 1 & -f_x d\theta & 0 \\ -f_z d\theta & f_x d\theta & 1 & 0 \\ 0 & 0 & 0 & 1 \end{bmatrix}. \tag{2.43}$$

Since $\Delta = Trans(dx, dy, dz)Rot(f, d\theta)$, it yields,

$$\Delta = \begin{bmatrix} 1 & 0 & 0 & dx \\ 0 & 1 & 0 & dy \\ 0 & 0 & 1 & dz \\ 0 & 0 & 0 & 1 \end{bmatrix} = \begin{bmatrix} 1 & -f_z d\theta & f_y d\theta & 0 \\ f_z d\theta & 1 & -f_x d\theta & 0 \\ -f_y d\theta & f_x d\theta & 1 & 0 \\ 0 & 0 & 0 & 1 \end{bmatrix} - \begin{bmatrix} 1 & 0 & 0 & 0 \\ 0 & 1 & 0 & 0 \\ 0 & 0 & 1 & 0 \\ 0 & 0 & 0 & 1 \end{bmatrix}$$

$$= \begin{bmatrix} 0 & -f_z d\theta & f_y d\theta & dx \\ f_z d\theta & 0 & -f_x d\theta & dy \\ -f_y d\theta & f_x d\theta & 0 & dz \\ 0 & 0 & 0 & 0 \end{bmatrix} \tag{2.44}$$

The differential rotation $d\theta$ about vector f is equivalent to the differential rotation with respect to the x, y and z-axis, namely δ_x, δ_y, and δ_z, respectively. Then $f_x d\theta = \delta_x, f_y d\theta = \delta_y, f_z d\theta = \delta_z$. Displace the above results into (2.44) yields:

$$\Delta = \begin{bmatrix} 0 & -\delta_z & \delta_y & dx \\ \delta_z & 0 & -\delta_x & dy \\ -\delta_y & \delta_x & 0 & dz \\ 0 & 0 & 0 & 1 \end{bmatrix}. \tag{2.45}$$

If $d = d_x i + d_y j + d_z k, \delta = \delta_x i + \delta_y j + \delta_z k$, then the differential motion vector of rigid body or coordinate frame can be expressed as follows:

$$\mathbf{D} = \begin{bmatrix} dx \; dy \; dz \; \delta_x \; \delta_y \; \delta_z \end{bmatrix}^T = \begin{bmatrix} d \\ \delta \end{bmatrix}. \tag{2.46}$$

The linear transformation between motion speed of manipulator and each joint can be defined as the Jacobian matrix of a robot. This Jacobian matrix represents the drive ratio of motion velocity from the space of joints to the space of end-effector. Assume the motion equation of manipulator

$$x = x(q) \tag{2.47}$$

represents the displacement relationship between the space of operation (end-effector) and the space of joints. Differentiating (2.47) with respect to time yields,

$$\dot{x} = J(q)\dot{q}, \tag{2.48}$$

where $\dot{x}$ is the generalized velocity of end-effector in operating space. $\dot{q}$ is the joint velocity. $J(q)$ is $6 \times n$ partial derivative matrix which is called Jacobian Matrix. The component in line i and column j is:

$$\mathbf{J}_{ij}(q) = \frac{\partial x_i(q)}{\partial q_j}, \quad i = 1, 2, \ldots, 6; \quad j = 1, 2, \ldots, n \tag{2.49}$$

From (2.49), it is observed that Jacobian Matrix $J(q)$ is a linear transformation from the velocity of joints space.

The generalized velocity $\dot{x}$ of rigid body or coordinate frame is a six-dimensional column vector composed of linear velocity v and angular velocity w.

$$\dot{x} = \begin{bmatrix} \mathbf{v} \\ \mathbf{w} \end{bmatrix} = \lim_{\Delta t \to 0} \frac{1}{\Delta t} \begin{bmatrix} d \\ \delta \end{bmatrix}. \tag{2.50}$$

Equation (2.50) can be rewritten as

$$\mathbf{D} = \begin{bmatrix} d \\ \delta \end{bmatrix} = \lim_{\Delta t \to 0} \dot{x} \Delta t. \tag{2.51}$$

Replace (2.48) into (2.50), one has:

$$\mathbf{D} = \lim_{\Delta t \to 0} J(q)\dot{q}\Delta t, \tag{2.52}$$

$$\mathbf{D} = J(q)dq. \tag{2.53}$$

For a robot with n joints, its Jacobian matrix is a $6 \times n$ matrix, in which the first three lines denote the transferring rate of end-effector's linear velocity, and the last three lines denote the transferring rate of end-effector's angular velocity. Jacobian matrix can be expressed as:

$$\begin{bmatrix} \mathbf{v} \\ \mathbf{w} \end{bmatrix} = \begin{bmatrix} J_{l1} & J_{l2} & \cdots & J_{ln} \\ J_{a1} & J_{a2} & \cdots & J_{an} \end{bmatrix} = \begin{bmatrix} \dot{q}_1 \\ \dot{q}_2 \\ \vdots \\ \dot{q}_n \end{bmatrix}. \tag{2.54}$$

The linear velocity and angular velocity of an end-effector can be expressed as the linear function of each joint velocity $\dot{q}$

$$v = J_{l1}\dot{q}_1 + J_{l2}\dot{q}_2 + \cdots + J_{ln}\dot{q}_n; \; \mathbf{w} = J_{a1}\dot{q}_1 + J_{a2}\dot{q}_2 + \cdots + J_{an}\dot{q}_n, \tag{2.55}$$

where J_{li} and J_{ai} means the linear velocity and angular velocity of end-effector resulted in joint i.

2.6 Conclusions

In this chapter, kinematics of robot manipulators is introduced, including the concept of reference coordinate frame, translational transformation, rotational transformation and homogeneous transformation, as well as the basic knowledge in robot kinematics, such as Euler angle, Denavit–Hartenberg representation, and Jacobian matrix of robot. These are the important knowledge in parallel robotic machine design.

Chapter 3
Architectures of Parallel Robotic Machine

3.1 Preamble

One of the objectives of this book is to find the most promising kinematic structures that can be used for machine tool design. Hence, some well-known principles are applied to investigate all the possibilities of structure in detail. A mechanism is defined as a kinematic chain with one of its components (link or joint) connected to the frame. A kinematic chain consists of a set of links, coupled by joints (cylindrical, planar, screw, prismatic, revolute, spherical, and Hooke) between adjacent links. In this chapter, a topological study of different combinations of kinematic chain structures are performed using a graph representation approach. The number of links and joints for the desired system and their interconnections, neglecting geometric details (link length and link shape), are described. The possible architectures that provide 5 degrees of freedom between the tool and the workpiece are generated. In Sect. 3.2, basic kinematic elements of mechanisms are introduced, and the classification of mechanisms is given based on the motion relation. In Sect. 3.3, the basic concept of the graph representation of a kinematic structure is addressed. Then, the Chebychev–Grübler–Kutzbach criterion is introduced in Sect. 3.4. A topological study of the kinematic structures is described in Sect. 3.5. Requirements for possible kinematic structures are set up. Furthermore, the structural representation of kinematic chains and architectures with consideration of parallel and hybrid cases is illustrated. In Sect. 3.6, a remark on the role of redundancy is given. A summary with discussion of related work is presented in Sect. 3.7.

3.2 Fundamentals of Mechanisms

3.2.1 Basic Kinematic Elements of Mechanisms

A mechanism is defined as a kinematic chain with one of its components (link or joint) connected to the frame. A kinematic chain consists of a set of links, coupled by joints between adjacent links.

D. Zhang, *Parallel Robotic Machine Tools*, DOI 10.1007/978-1-4419-1117-9_3,

3.2.1.1 Prismatic Joint (P, also called sliders)

A prismatic joint allows two components to produce relative displacement along the common axis. The included angle between the two components is a constant value, called deflection angle. The displacement and deflection angle describe the spatial relative relationship of the two components, which forms a prismatic joint. A prismatic joint is a one degree-of-freedom kinematic pair, which provides single-axis sliding function, and it can be used in places such as hydraulic and pneumatic cylinders. The CAD model of a prismatic joint is shown in Fig. 3.1.

3.2.1.2 Revolute Joint (R, also called pin joint or hinge joint)

A revolute joint allows two components produce relative rotation along the joint axis. The vertical dimension between the two components, is a constant value called offset distance. The vertical dimension and offset distance describe the spatial relative relationship of the two components which forms a revolute joint. A revolute joint, as a one degree-of-freedom kinematic pair, provides single-axis rotation function. Revolute joints is the most commonly found joint in industrial and research robots, and it can be found in many classic applications, such as door hinges, folding mechanisms, and other uniaxial rotation devices. The CAD model of a revolute joint is shown in Fig. 3.2.

Fig. 3.1 The CAD model of prismatic joint

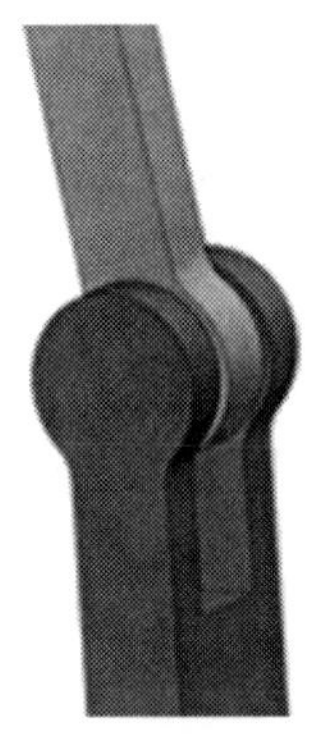

Fig. 3.2 The CAD model of revolute joint

3.2.1.3 Hooke Joint (H, also called universal joint, Cardan joint or Hardy-Spicer joint)

Hooke joint allows two components to produce two degree-of-freedom relative independent rotation along two perpendicular axes. Generally, a Hooke joint is equivalent to two revolute joints whose axes must be completely perpendicular, namely H = RR. The CAD model of a Hooke joint is shown in Fig. 3.3.

3.2.1.4 Spherical Joint (S, also called ball-in-socket joint)

A spherical joint allows one element to rotate freely in three dimensions with respect to the other about the center of a sphere. The sense of each rotational degree-of-freedom is defined by the right-hand rule, and the three rotations together form a right-hand system. The relative pose of two components can be confirmed by three Euler angles, ϕ (rotate along the original z-axis), θ (rotate along the new x-axis), and φ (rotate along the new z-axis). A spherical joint is kinematically equivalent to three intersecting revolute joints. The CAD model of a Hooke joint is shown in Fig. 3.4.

Fig. 3.3 The CAD model of Hooke joint

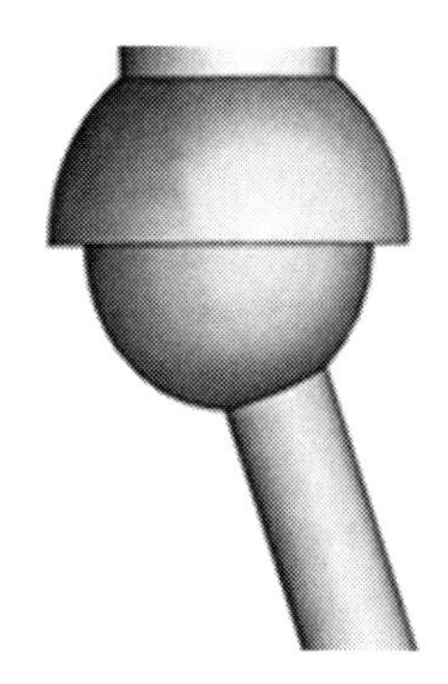

Fig. 3.4 The CAD model of Spherical joint

3.2.2 Classification of Mechanisms

Mechanisms can be divided into planar mechanisms and spatial mechanisms, according to the relative motion of the rigid bodies.

Serial mechanisms have been extensively studied in terms of their design, kinematic and dynamic modeling, and control by many researchers. When properly designed, the serial structure has the benefit of possessing a large workspace volume in comparison to the physical size of the mechanism. Since serial mechanisms only have one open kinematic chain, this means that the serial mechanisms only have one possibility in architecture.

Parallel mechanism is closed-loop mechanism in which the end-effector is connected to the base by at least two independent kinematic chains [106]. This can be further divided into fully-parallel and hybrid mechanism. Fully-parallel mechanism is the one with an n-DOF end-effector connected to the base by n independent kinematic chains, each having a single actuated joint. The hybrid one has the combination of serial and parallel mechanisms.

Because their errors are averaged instead of added cumulatively, parallel robots are more accurate than serial robots. First, since the moving platform of parallel mechanism is supported by several kinematic chains, the system stiffness of the end-effector is largely improved. Furthermore, this also strengthens the structural stability. Contrarily, serial mechanism usually is a single-arm structure. To some extent, a large number of motors increase the burden of the end-effector and affect the structural stability of serial mechanism. Second, the specific configuration of parallel mechanism makes it have obvious advantages in the abilities of reconfiguration, restoration, and payload. Third, the error of the end-effector of serial mechanism will be accumulated and amplified based on each joint error; contrarily, the error of parallel mechanism is smaller and its accuracy is higher. Fourth, the actuators of serial robot usually are located on the end of each rod end. It will increase the inertia and exacerbate the transfer ability of system. For parallel mechanisms, the actuators can be located on the base to decrease the motion load.

With the development of the theory of advanced spatial mechanism and the technology of robotics, parallel robotic machines have been an important branch of robotic technology. Furthermore, the research activities of the theories and applications of parallel robots are becoming increasing. Nevertheless, many scholars have done intensive investigations on the dimension synthesis, kinematics, dynamics, workspace, and singularity of parallel mechanisms, most of the existing work regarding parallel mechanisms was built upon the concept of traditional Gough-Stewart mechanism type. Because of the opposition and unitarian of serial mechanisms and parallel mechanisms in philosophy, the hybrid mechanisms can be built through the combination of parallel and serial mechanisms and play an important role in some specific application background.

The number of independent coordinates to completely determine the location of an object in space can be called the degree-of-freedom of the object. In the Cartesian coordinate system, three independent coordinates *(xyz)* must be used to confirm the position of a particle with random motion. Thus, a free particle has three translational degree-of-freedom. Likelihood, the free motion of a rigid body in three dimensional spaces can be decomposed into the translational motion of its barycenter and the rotational motion with respect to the axis of barycenter. Therefore, a rigid body with random motion totally has six degree-of-freedom: three translations *(xyz)* to measure its position and three rotations (ϕ, θ, φ) to measure its pose. The definition of degree-of-freedom for parallel mechanism, similar to the motion of a rigid body in space, is the sum of independent translational degree-of-freedom and independent rotational degree-of-freedom of end-effector (attached to the moving platform) with respect to a fixed coordinate system. Generally, the fixed coordinate system is attached to the fixed base. Sometimes, the end-effector can only produce motion in a plane. Since a rigid body has three degree-of-freedom in a plane: two translations in *(xy)* and one autogiration, the planar parallel mechanism has at most three degrees of freedom.

A brief introduction of parallel mechanism based on the classification of space dimension and degree-of-freedom is given as follows.

1. Planar two degrees of freedom parallel mechanism
   Cervantes [30] proposed a simplified approach which allowed the generation of the workspace of a complete class of 2-dof manipulators with the type of RPRPR.
   Tensegrity structure is combined by a group of continuous/discontinuous draw bar to form a self-stress, self-supporting reticulated linkage structure. Arsenault [13] designed a planar two degree-of-freedom modular parallel mechanism based on the principle of tensegrity.
2. Planar three degrees of freedom parallel mechanism
   Zhang [179] proposed a planar three degree-of-freedom parallel mechanism with redundant actuation. Since specific driven redundancy method is adopted, the closed-form solution for the forward kinematics was derived.
3. Spatial three degrees of freedom parallel mechanism
   Clavel's delta parallel robot [128] is the classic case of spatial three degree-of-freedom mechanism. The parallelograms are adopted in three symmetrical legs to improve the dynamics performance. Delta robot has 50 gravitational accelerations in the environment of laboratory. Even in the industrial field, it still has 12 gravitational accelerations. In the process of three degree-of-freedom linear motion, the leg in the parallelogram must always keep parallel to its opposite side.
4. Spatial four degrees of freedom parallel mechanism
   Alvarado [50] proposed a four degree-of-freedom CPS+PS+HPS parallel mechanism with three legs. The numerical analysis results showed the efficiency of screw theory when dealing with the issues of kinematics and singularity of simple parallel mechanism.
   Lu [100] analyzed the kinematics and active/passive force of a four degree-of-freedom 3SPU+UPR mechanism with three rotations and one translation.
   Inspired by Clavel's Delta robot, Olivier et al. [39] developed the prototype of H4 robot using parallelogram mechanism.

Kong et al. [85] proposed a general approach for the type synthesis of a class of parallel mechanisms based on screw theory. The common ground of these mechanisms is that they have completely same branch chain, 3T1R.

5. Spatial five degrees of freedom parallel mechanism
   Alizade [8] discussed a kind of five degree-of-freedom 4UPS+UPU asymmetry parallel mechanism with three rotations and two one translations.
6. Spatial six degrees of freedom parallel mechanism
   Gough-Stewart platform is the original of spatial six degree-of-freedom parallel mechanism.

3.3 Graph Representation of Kinematic Structures

A kinematic chain can be described as a set of rigid bodies attached to each other by kinematic pairs, resulting in a mechanical network containing joints and links [56]. A kinematic structure represents the kinematic chain without considering the detailed geometric, kinematic, and functional properties. The range of kinematic structures given particular constraints on the number and type of joints and links can be examined exhaustively. This range represents a set of logical possibilities for design of a particular type of mechanism. This set is a framework in which designs are to be realized.

A systematic method of enumerating all the possible kinematic chains – kinematic architectures – is needed to meet the required degrees of freedom, i.e. 3-dof, 4-dof, and 5-dof. There were several methods reported in the literature: Hunt [76] used the theory of screw systems to enumerate parallel mechanisms exhaustively; Earl et al. [44] proposed a network approach, which enables consideration of two or more structures into another one. A graph representation will be introduced in this chapter.

Graph theory is a field of applied mathematics [67], which provides a useful abstraction for the analysis and classification of the topology of kinematic chains, and it offers a systematic way of representing the topology of complex kinematic chains. The graph of a kinematic chain consists of a diagram where each link is represented by a point and each joint by a line. Thus, the graph representation of a kinematic chain will take the form of a collection of points connected by lines. The graph representation of kinematic chains has been used by many researchers [16, 56, 146, 180, 181].

3.4 Design Criteria

The degree of freedom (or mobility) of a kinematic chain [76] can be defined as the minimum number of independent variables necessary to specify the location of all links in the chain relative to a reference link. The choice of the reference link does not affect the resulting mobility. A preliminary evaluation of the mobility of a kinematic chain can be found from the Chebychev-Grübler-Kutzbach formula.

$$l = d(n - g - 1) + \sum_{i=1}^{g} f_i, \tag{3.1}$$

where l is the degree of freedom of the kinematic chain, d is the degree of freedom of each unconstrained individual body (6 for the spatial case, 3 for the planar case) [77]; n is the number of rigid bodies or links in the chain; g, the number of joints; and f_i, the number of degrees of freedom allowed by the ith joint.

For example, to design a 5-DOF parallel robotic machine, the possibility of parallel mechanisms can be investigated for the combinations of dofs in (5,0), (4,1), and (3,2). The workpiece can be fixed (0-dof), or move along one axis (1-dof) or move along the X and Y axes (2-dof) or rotate about one or two axes. Hence, one will consider the possibilities of parallel mechanisms with 5-dof, 4-dof, 3-dof, and 2-dof; besides, the case with 6-dof is taken as an option with redundancy. The detail is shown as follows.

1. DOF distributions for each leg
   For a given parallel platform, we can always make the following assumptions:

   number of known bodies = 2 (platform and base),
   number of parallel legs = L, and
   degree of freedom of the ith leg = f_{l_i},

   then one can rewrite (10.1) as

$$\begin{aligned} l &= 6\left[2 + \sum_{i=1}^{L}(f_{l_i} - 1) - \sum_{i=1}^{L} f_{l_i} - 1\right] + \sum_{i=1}^{g} f_i \\ &= 6 - 6L + \sum_{i=1}^{g} f_i. \end{aligned} \tag{3.2}$$

   From this equation, it is apparent that there exist thousands of possibilities for 5-dof or less than 5-dof cases. Hence, some constraints introduced and are specified as follows:

   - From the viewpoint of fully-parallel mechanism, the maximum number of parallel legs are kept equal to the degree of freedom of the mechanism, thus to guarantee the possibility of installing one actuator in each leg, one has

$$L \leq l. \tag{3.3}$$

   - Although two-leg spatial parallel mechanisms are of little direct use independently, they are useful to constructing "Hybrid" mechanisms, the minimum number of the leg is given by

$$L \geq 2. \tag{3.4}$$

Table 3.1 The possible degree-of-freedom distribution for each leg

Degree of freedom	Number of legs	f_{l_1}	f_{l_2}	f_{l_3}	f_{l_4}	f_{l_5}	f_{l_6}
l = 2	L = 2	2	6				
		3	5				
		4	4				
l = 3	L = 2	3	6				
		4	5				
	L = 3	3	6	6			
		4	5	6			
		5	5	5			
l = 4	L = 2	4	6				
		5	5				
	L = 3	4	6	6			
		5	5	6			
	L = 4	4	6	6	6		
		5	5	6	6		
l = 5	L = 2	5	6				
	L = 3	5	6	6			
	L = 4	5	6	6	6		
	L = 5	5	6	6	6	6	
l = 6	L = 2	6	6				
	L = 3	6	6	6			
	L = 4	6	6	6	6		
	L = 5	6	6	6	6	6	
	L = 6	6	6	6	6	6	6

On the basis of the constraints represented by (3.3) and (3.4), and one can enumerate the possible dofs distributions as in Table 3.1. It is noted that these are the basic combinations for different architectures, and one can remove or add legs which have 6-dof for symmetric purpose in any of the basic structures at ease.

3.5 Case Study: Five Degrees of Freedom Parallel Robotic Machine

Since both the tool and the workpiece can be actuated independently and that 5-DOF are required for manufacturing tasks, the possible combinations of 5 dofs are: (5,0), (4,1), and (3,2) as indicated in previous section. For each of these combinations, the kinematic chains involved may lead to several possibilities (serial, parallel, or hybrid). The followings are the details for this enumeration process.

3.5.1 Serial Mechanisms

The serial mechanisms have many drawbacks. Because of the serial nature of actuation and transmission, related masses must be mounted distal to the base of the mechanism leading to a small ratio of payload over machine mass, poor dynamic performance in terms of acceleration capability, and poor system stiffness presented at the end-effector. Since a lower axis has to carry both the loads (in all directions) and the weights of all its upper axes, dynamic behaviors of the lower axes will be poor, especially to machine tools which carry high loads. In addition, the serial structure leads to joint errors being additive, and combined with the inherent low system stiffness, this leads to poor accuracy at the end-effector. Thus, the drawbacks in their structures limit the performance.

3.5.2 Parallel Mechanisms

Among the three possibilities (serial, parallel, and hybrid), the parallel mechanisms are the basic and the most important ones in building all the possible architectures, because of the disadvantages of the serial mechanisms. The hybrid mechanisms will be built through the combination of parallel mechanisms.

1. Possible Structures
   The variables for combining different kinds of architectures are mainly decided by (1) leg length; (2) position of the base points; or (3) both the leg length and position of the base points.

   (a) Possible Legs
   On the basis of the required DOF distributions for each leg, one can find different kinds of legs to meet the requirement through the combination of different joints such as spherical joint (with 3-dof), Hooke joint (with 2-dof), revolute joint (with 1-dof) and prismatic joint (with 1-dof). One can combine them to meet the dof requirements for each leg shown in Table 3.2, where

   S: spherical joint
   R: revolute joint
   H: Hooke joint
   P: prismatic joint

   Table 3.3 shows all the possible legs with a different degree-of-freedom.

   (b) Vertex structures
   From the literature related to the Stewart platform, various architectures have been developed or proposed for the platform mechanisms, such as 3-6, 4-4, 4-5, and 4-6 (the numbers of vertices in the mobile and base plates) platforms [33,48,63,99,169]. Since two spherical joints can be combined to one concentric spherical joint, one can obtain two types of vertices for parallel mechanisms as shown in Fig. 3.5.

Table 3.2 Possible joint combinations for different degrees of freedom

Number of possibilities	DOFs = 2	DOFs = 3	DOFs = 4	DOFs = 5	DOFs = 6
1	2R	1R2P	1S1P	1S2R	2S
2	2P	2R1P	1S1R	1S2P	1S1H1P
3	1R1P	3R	1R3P	1S1R1P	1S1H1R
4	1H	3P	2R2P	1S1H	1S3R
5		1H1R	3R1P	1H3R	1S3P
6		1H1P	4R	1H2R1P	1S2R1P
7		1S	4P	1H1R2P	1S1R2P
8			1H2R	1H3P	1H4R
9			1H2P	5R	1H3R1P
10			1H1R1P	4R1P	1H2R2P
11				3R2P	1H1R3P
12				2R3P	1H4P
13				1R4P	6R
14				5P	5R1P
15					4R2P
16					3R3P
17					2R4P
18					1R5P
19					6P

Table 3.3 Possible leg types with different degrees of freedom

Possible numbers	DOFs = 2	DOFs = 3	DOFs = 4	DOFs = 5	DOFs = 6
1	2R	1S	2R1H	1H2R1P	1S2R1P
2	1R1P	2R1P	1H1R1P	2H1R	1S1H1P
3	1H	1R2P	1S1P	2H1P	1S1H1R
4		3R	1S1R	1S1R1P[1]	1S1R1P
5		3P	2R2P	1S2R[1]	1S1H1P
6		1H1R	1R3P	1S2P	2S
7		1H1P	3R1P	1H3P	1S3P
8			4R	1H1R2P	1S3R
9			4P	1H3R	1H2R2P
10			1H2P	4R1P	1H3R1P
11				5R	1H4R
12				5P	6R
13				1R4P	6P
14				3R2P	1H4P
15				2R3P	3R3P
16					1H1R3P
17					5R1P
18					1R5P
19					2R4P
20					4R2P
Total possibilities	3	7	10	15	20

[1]They are only suitable for those with identical legs, e.g., 3-DOF mechanism.

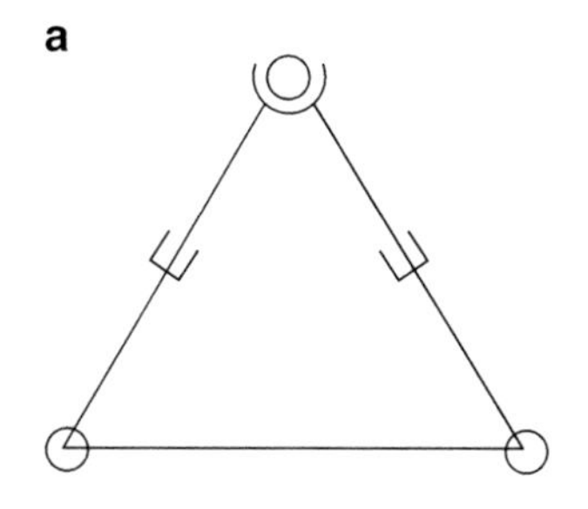

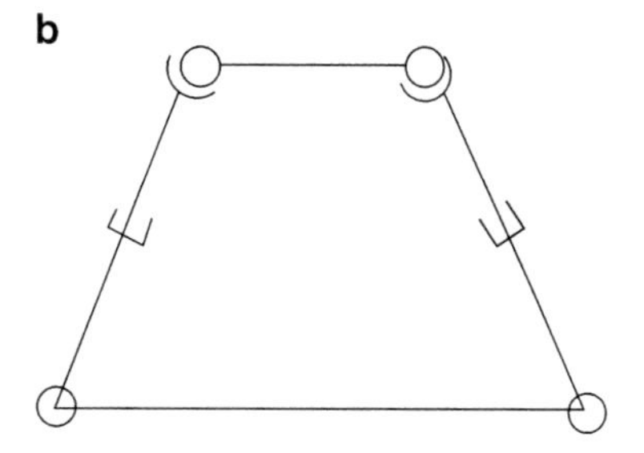

Fig. 3.5 Two types of vertex structures

Fig. 3.6 Possible architectures with 3 legs

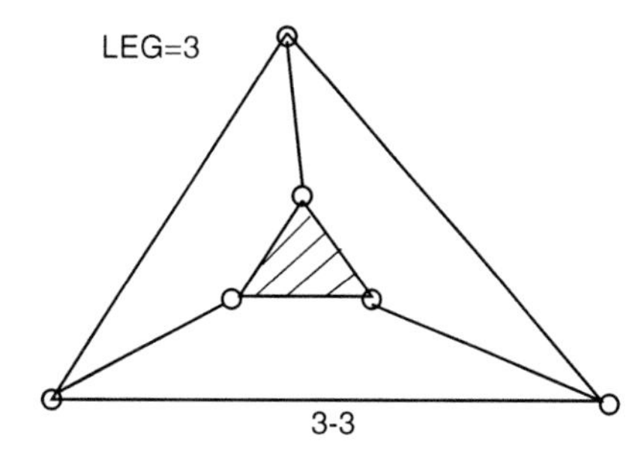

On the basis of these two vertex structures, various types of parallel mechanism structures can be obtained through different arrangements of the joints on the base and mobile platforms.

(c) Platform structures
Once the type of vertex structure is decided, one can obtain the platform structure according to the number of vertices.

2. Possible architectures for parallel mechanisms
On the basis of the above analysis, one can assemble all the possible architectures as shown in Figs. 3.6 – 3.9.
3. The most promising architectures
As listed in Tables 3.1 and 3.2, although we have already given constraints to DOF distributions for each leg, there are still lots of possible combinations for parallel mechanisms which meet the machine tool's DOF requirement, *e.g.*, for DOFs = 3, from Table 3.1, there are 3 possible combinations of legs with degree-of-freedom of 3, 4, 5, and 6. Meanwhile, from Table 3.2, there are 7, 10, 14, and 19 possible combinations for legs with dofs of 3, 4, 5, and 6, thus we still have many architectures through the permutation and combination. To find the most promising architectures, the criteria for selection of joints and legs are given as follows

(a) Proper number and type of DOFs
In order to ensure the required motions (i.e., 5-dof between the tool and the workpiece) in Table 3.4, the DOFs distribution numbers and the type of

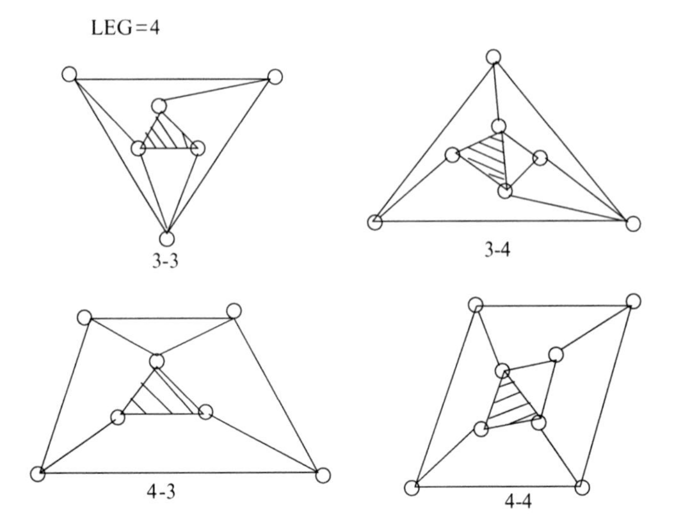

Fig. 3.7 Possible architectures with 4 legs

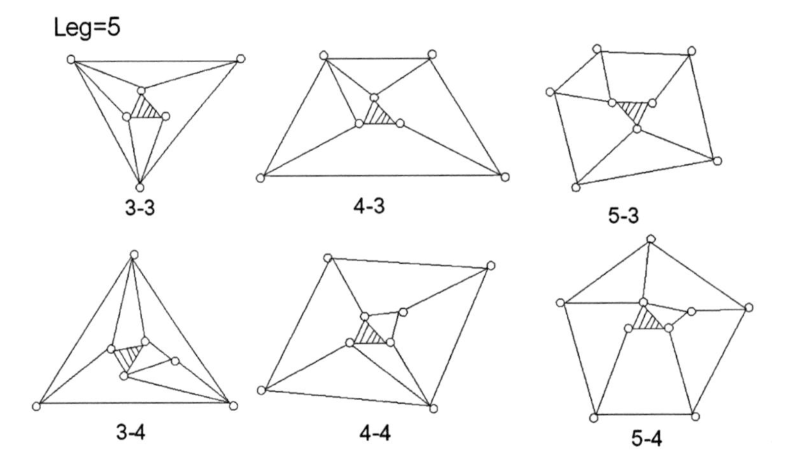

Fig. 3.8 Possible architectures with 5 legs

motions for each leg should be properly arranged. Each leg can be facilitated with spherical, prismatic, Hooke and revolute joints.

(b) Simplicity and practicability
The legs used in machine tools must be simple and practical. For the sake of the simplicity and dexterity of mechanism, we prefer to use "spherical"pairs as the joints between link and platform for those legs with more than 3 dofs. Since the serially connected revolute joints easily lead to "Singularity" and the "manufacturability" is difficult, so we abandon to use of more than 2 revolute joints connected in series.

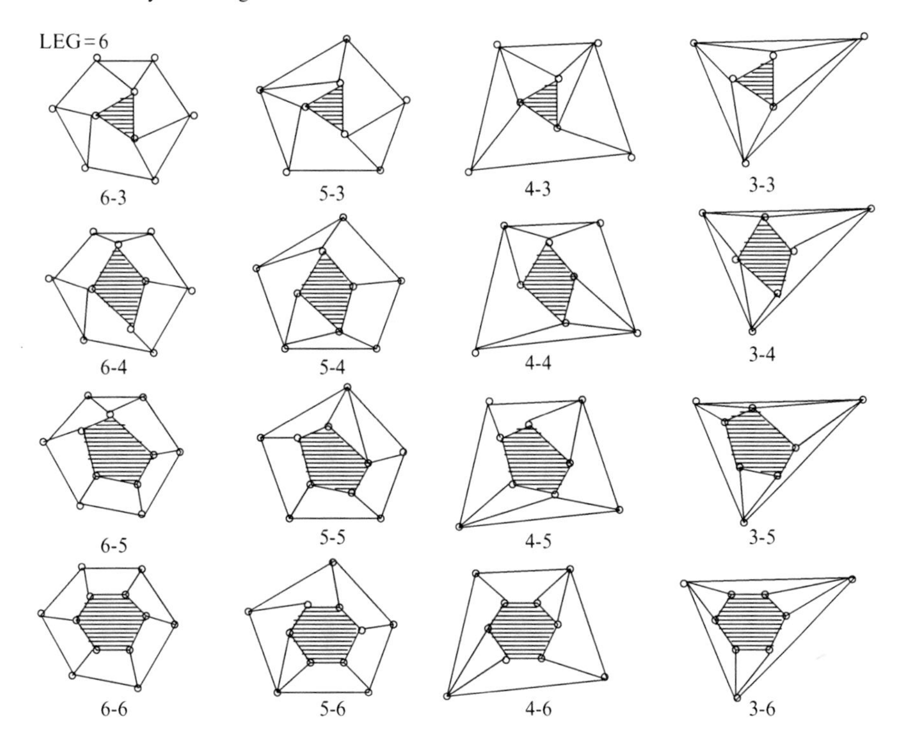

Fig. 3.9 Possible architectures with 6 legs

Table 3.4 The possible motion distributions for required 5-dof between the tool and the workpiece

DOFs (machine tools)	Motion of workpiece	Motion of machine tool
$l = 3$	X, Y: translation	Z: translation, X, Y: rotation
	X, Y: rotation	X, Y, Z: translation
	combination of R & T	X, Y, Z: combination of R & T
$l = 4$	X (or Y) translation	X, (or Y), Z: translation; X, Y: rotation
	X, (or Y): rotation	X, Y, Z: translation; X, (or Y): rotation
	combination of R & T	X, Y, Z: combination of R & T
$l = 5$	fixed	X, Y, Z: translation; X, Y: rotation

(c) Elimination of passive prismatic joints
Because it is difficult to control passive prismatic joints, in order to avoid the existence of passive prismatic joints, we specify

$$\text{Number of actuators} \geq \text{Number of prismatic joints} \tag{3.5}$$

Meanwhile, as we desire to put the actuators at the base of each link, therefore at most one prismatic joint can be used for each leg.

Table 3.5 Possible leg types with different degrees of freedom

Possible numbers	DOFs = 2	DOFs = 3	DOFs = 4	DOFs = 5	DOFs = 6
1	2R	1S	2R1H	1H2R1P	1S2R1P
2	1R1P	2R1P	1H1R1P	2H1R	1S1H1P
3	1H			2H1P	1S1H1R
4				1S1R1P[1]	
5				1S2R[1]	
6				1S1H[1]	
Total possibilities	3	2	2	6	3

[1]They are only suitable for those with identical legs, e.g., 3-DOF mechanism.

Table 3.6 The possible architectures

Degree of freedom	Number of legs	f_{l_1}	f_{l_2}	f_{l_3}	f_{l_4}	f_{l_5}	f_{l_6}	Possible architectures	Possible architectures with identical dof structure	$L = l$
$l = 2$	L = 2	2	6					9	9	9
		3	5					12	12	12
		4	4					3	2	2
$l = 3$	L = 2	3	6					6	6	
		4	5					12	12	
	L = 3	3	6	6				12	6	6
		4	5	6				36	36	36
		5	5	5				56	6	6
$l = 4$	L = 2	4	6					6	6	
		5	5					6	3	
	L = 3	4	6	6				12	6	
		5	5	6				18	9	
	L = 4	4	6	6	6			20	6	6
		5	5	6	6			36	9	9
$l = 5$	L = 2	5	6					9	9	
	L = 3	5	6	6				18	9	
	L = 4	5	6	6	6			30	9	
	L = 5	5	6	6	6	6		45	9	9
$l = 6$	L = 2	6	6					9	3	
	L = 3	6	6	6				10	3	
	L = 4	6	6	6	6			15	3	
	L = 5	6	6	6	6	6		21	3	
	L = 6	6	6	6	6	6	6	28	3	3
Total								429	179	98

(d) Elimination of the rotation around the Z axis

Since the rotation around the Z axis is not needed, we can introduce a n-dof passive leg into the mechanism to reach the desired motion. "Spherical joint" on the movable platform will be replaced by "Hooke joint" + "Prismatic joint" or "Hooke joint" + "Revolute joint" so as to constrain the rotation around the Z axis. The passive constraining leg will be put in the center of the platform to minimize the torque and force. Since the external loads on the platform will induce a bending and/or torsion in the passive leg, its mechanical design is a very important issue, which can be addressed using the kinetostatic model later. In this case, the actuators are put in each of the identical legs and leave the special one (different DOFs) as the passive link since its structure in design size is larger than the other legs to sustain the large wrench.

(e) Structure of the mechanisms
The study is based on fully-parallel mechanisms, but one can add legs (with 6-dof) to keep the structure symmetric. For the shape of the platforms, one should avoid the use of regular polygon, since it may lead to geometry singularity.

Based on the discussion above, we eliminate some of the impractical joint combinations and obtain the prospective ones as shown in Table 3.5.

Through the combinations of the possibilities, we obtain the number of the most promising possible architectures shown in Table 3.6. When $L = l$, we obtain a fully-parallel mechanism.

3.5.3 Hybrid Mechanisms

A hybrid (serial-parallel) mechanism is a combination of serial and parallel mechanisms. It comprises two parallel actuated mechanisms connected in series, one of them is the upper stage, the other is the lower stage, and the moving platform of the lower stage is the base platform of the upper stage. This special structure results in a mechanism with the attributes of both. It provides a balance between exclusively serial and parallel mechanisms and better dexterity. It can even improve the ratio of workspace to architecture size and the accuracy.

To meet the required 5-dof motion, 2-dof and 3-dof parallel mechanisms are chosen to construct the "Hybrid" mechanisms. Since the upper stage is connected with the end-effector, and it requires high stiffness, so a 3-dof parallel mechanism is considered as the upper stage while a 2-dof parallel stage is taken as the lower stage.

For a 2-dof parallel mechanism – the lower stage of hybrid mechanism – both planar and spatial parallel mechanisms can be considered. Referring to (3.2), for planar mechanisms ($d = 3$), then one has

$$\begin{aligned} l &= 3\left[2 + \sum_{i=1}^{L}(f_{l_i} - 1) - \sum_{i=1}^{L} f_{l_i} - 1\right] + \sum_{i=1}^{g} f_i \\ &= 3 - 3L + \sum_{i=1}^{g} f_i \end{aligned} \tag{3.6}$$

Table 3.7 The possible degree-of-freedom distribution for planar mechanisms

Degree of freedom	Number of legs	f_{l_1}	f_{l_2}	f_{l_3}
l = 2	L = 2	2	3	
l = 3	L = 3	3	3	3
		2	3	4

Therefore, the possible DOFs distribution for planar mechanisms can be found in Table 3.7.

The hybrid motions (5-dof) can be arranged as follows:

- Upper stage: *X*, *Y* axes rotation, *Z* axis translation; lower stage: X, Y axes translation
  One can realize this motion through either the combination of 3SPR as upper stage and "Linear motion components" (LM) as the lower stage (special case) or the combination of 3SPR as upper stage and 3RRR planar parallel mechanism as the lower stage.
- Upper stage: *X*, *Y* axes translation, *Z* axis translation; lower stage: *X*, *Y* axes rotation
  One can realize this motion through the combination of 3SRR as upper stage and 2-dof spherical parallel mechanism as the lower stage. Because of the complexity in manufacturing spherical parallel mechanisms, low stiffness, low precision, and small workspace, we discard spherical parallel mechanisms in our research.

The "Hybrid" mechanisms can also be implemented in an alternative way, i.e., using positioning head (wrist) for machine tools design, this will be described in the next section.

3.6 Redundancy

The main purpose of adopting redundancy is to improve reliability and dexterity. To make the parallel kinematic machines capable of arbitrarily positioning and orienting the end-effector in a three-dimensional workspace, redundancy factor may be considered. In this book, only 3-dof, 4-dof, 5-dof, and 6-dof spatial parallel mechanisms are discussed. Generally, all these types of mechanisms are used for base platform, one can select a positioning head (wrist) with 1-dof, 2-dof, or 3-dof in conjunction with the base platform. This constructs a hybrid mechanism and it will lead to some redundant cases.

3.7 Conclusions

The kinematic structures used for 5-dof or less than 5-dof machine tools design with their underlying design principles have been made more explicit through the discussion and enumeration in this chapter. From the results obtained, it can be seen that both the tool and the workpiece can be actuated independently and that 5-dof is required for manufacturing tasks, the possible combinations of degree-of-freedom are: (5,0), (4,1), and (3,2). Moreover, for each of these combinations, the kinematic chains involved lead to several possibilities (serial, parallel, or hybrid) and additionally, redundancy is taken as an option. Finally, a detailed list of possible topologies has been obtained and the most promising architectures are pointed out under the design criteria.

Chapter 4
Planar Parallel Robotic Machine Design

4.1 Preamble

Parallel kinematic machines with their unique characteristics of high stiffness (their actuators bear no moment loads but act in a simple tension or compression) and high speeds and feeds (high stiffness allows higher machining speeds and feeds while providing the desired precision, surface finish, and tool life), combined with versatile contouring capabilities have made parallel mechanisms the best candidates for the machine tool industry to advance machining performance. It is noted that the stiffness is the most important factor in machine tool design since it affects the precision of machining. Therefore, to build and study a general stiffness model is a very important task for machine tool design. In this chapter, we will build a general stiffness model through the approach of kinematic and static equations. The objective of this model is to provide an understanding of how the stiffness of the mechanism changes as a function of its position and as a function of the characteristics of its components. This can be accomplished using stiffness mapping.

There are two methods to build mechanism stiffness models [170]. Among them, the method which relies on the calculation of the parallel mechanism's Jacobian matrix is adopted in this book.

It will be shown that the stiffness of a parallel mechanism is dependent on the joint's stiffness, the leg's structure and material, the platform and base stiffness, the geometry of the structure, the topology of the structure, and the end-effector position and orientation.

Since stiffness is the force corresponding to coordinate i required to produce a unit displacement of coordinate j, the stiffness of a parallel mechanism at a given point of its workspace can be characterized by its stiffness matrix. This matrix relates the forces and torques applied at the gripper link in Cartesian space to the corresponding linear and angular Cartesian displacements. It can be obtained using kinematic and static equations. The parallel mechanisms considered here are such that the velocity relationship can be written as in (4.1),

$$\dot{\theta} = \mathbf{J}\dot{\mathbf{x}}, \tag{4.1}$$

D. Zhang, *Parallel Robotic Machine Tools*, DOI 10.1007/978-1-4419-1117-9_4,

where $\dot{\theta}$ is the vector of joint rates and $\dot{\mathbf{x}}$ is the vector of Cartesian rates – a six-dimensional twist vector containing the velocity of a point on the platform and its angular velocity. Matrix $\mathbf{J}$ is usually termed Jacobian matrix, and it is the mapping from the Cartesian velocity vector to the joint velocity vector. From (4.1), one can conclude that

$$\delta\theta = \mathbf{J}\delta\mathbf{x}, \tag{4.2}$$

where $\delta\theta$ and $\delta\mathbf{x}$ represent joint and Cartesian infinitesimal displacements, respectively. Then, one can get the stiffness of this mechanism using the principle of kinematic/static duality. The forces and moments applied at the gripper under static conditions are related to the forces or moments required at the actuators to maintain the equilibrium by the transpose of the Jacobian matrix $\mathbf{J}$. This is also true for parallel mechanism [**?**], and one can then write

$$\mathbf{F} = \mathbf{J}^T\mathbf{f}, \tag{4.3}$$

where $\mathbf{f}$ is the vector of actuator forces or torques, and $\mathbf{F}$ is the generalized vector of Cartesian forces and torques at the gripper link, which is also called the wrench acting at this link [14, 165]. The actuator forces and displacements can be related by Hooke's law, one has

$$\mathbf{f} = \mathbf{K}_J\delta\theta \tag{4.4}$$

with $\mathbf{K}_{\mathrm{J}} = \mathrm{diag}[k_1, \ldots, k_n]$, where each of the actuators in the parallel mechanism is modeled as an elastic component, $\mathbf{K}_{\mathrm{J}}$ is the joint stiffness matrix of the parallel mechanism, k_i is a scalar representing the joint stiffness of each actuator, which is modeled as linear spring, and the ith component of vector $\mathbf{f}$, noted f_i is the force or torque acting at the ith actuator. Substituting (4.2) into (4.4), one obtains

$$\mathbf{f} = \mathbf{K}_{\mathrm{J}}\mathbf{J}\delta\mathbf{x}. \tag{4.5}$$

Then, substituting (4.5) into (4.3), yields

$$\mathbf{F} = \mathbf{J}^{\mathrm{T}}\mathbf{K}_{\mathrm{J}}\mathbf{J}\delta\mathbf{x}. \tag{4.6}$$

Hence, $\mathbf{K}_{\mathrm{C}}$, the stiffness matrix of the mechanism in the Cartesian space is then given by the following expression

$$\mathbf{K}_{\mathrm{C}} = \mathbf{J}^{\mathrm{T}}\mathbf{K}_J\mathbf{J}. \tag{4.7}$$

Particularly, in the case for which all the actuators have the same stiffnesses, i.e., $k_1 = k_2 = \cdots = k_n$, then (4.7) will be reduced to

$$\mathbf{K} = k\mathbf{J}^{\mathrm{T}}\mathbf{J}, \tag{4.8}$$

which is the equation given in [57].

The stiffness matrix is a positive semidefinite symmetric matrix whose eigenvalues represent the coefficients of stiffness in the principal directions, which are given by the eigenvectors. These directions are in fact represented by twist vectors, i.e., generalized velocity vectors. Moreover, the square root of the ratio of the smallest eigenvalue to the largest one gives the reciprocal of the condition number κ of the Jacobian matrix [83], which is a measure of the dexterity of the mechanism [56]. It can be written as

$$\frac{1}{\kappa} = \sqrt{\frac{\lambda_{\min}}{\lambda_{\max}}}, \tag{4.9}$$

where $\lambda_{\min}$ and $\lambda_{\max}$ are the smallest and largest eigenvalues of the stiffness matrix, respectively.

From (4.7), it is clear that if the Jacobian matrix of a mechanism $\mathbf{J}$ is singular, then obviously, the stiffness matrix of the mechanism, $\mathbf{J}^T\mathbf{K}_J\mathbf{J}$ is also singular, thus the mechanism loses stiffness, there is no precision also for the mechanism. Hence, one can study the precision of machine tools through their stiffness model, and then find the most suitable designs.

The flexibilities included in the model can be classified in two types [36] 1) the flexibilities at the joints and 2) the flexibilities of the links. Hence, the complete lumped model should include the following three submodels:

- The *Denavit–Hartenberg model* which defines the nominal geometry of each of the kinematic chains of the mechanism, the kinematics described by the Denavit–Hartenberg matrix are straightforward and systematic for mechanisms with rigid links. They are also effective for mechanisms with flexible links
- *A lumped joint model* which is defined in Table 4.1
- *An equivalent beam model* at each link which accounts for the deformations of the link caused by the external forces and torques

In order to simplify the model of the stiffness, link stiffnesses will be lumped into local compliant elements (spring) located at the joints. This is justified by the fact that no dynamics is included in the model (it is purely kinematic) and that limited numerical accuracy is acceptable. Indeed, the objective of this study is to obtain engineering values for the stiffness and to determine which areas of the workspace lead to better stiffness properties.

Physically, the bending deformation in joints is presented in different ways. In the planar case, the unactuated revolute joint does not induce any bending whereas in the spatial case, a bending is presented in a direction perpendicular to the joint. Hence, it is necessary to establish a lumped joint model for each possible case. In the lumped joint model, deformations caused by link flexibility can be considered as virtual joints fixed at this point; the details are given in [62] and Table 4.1.

A linear beam is shown in Fig. 4.1, where F is the external force, E the elastic modulus, L the length of the beam, and I is the section moment of inertia of the beam. In a lumped model, the flexible beam will be replaced by a rigid beam mounted on a pivot plus a torsional spring located at the joint, as illustrated in Fig. 4.1b. The objective is to determine the equivalent torsional spring stiffness that

Table 4.1 Lumped joint models for planar system

Joint type	If actuated, the equivalent model	If unactuated, the equivalent model
Revolute	2 Torsional springs 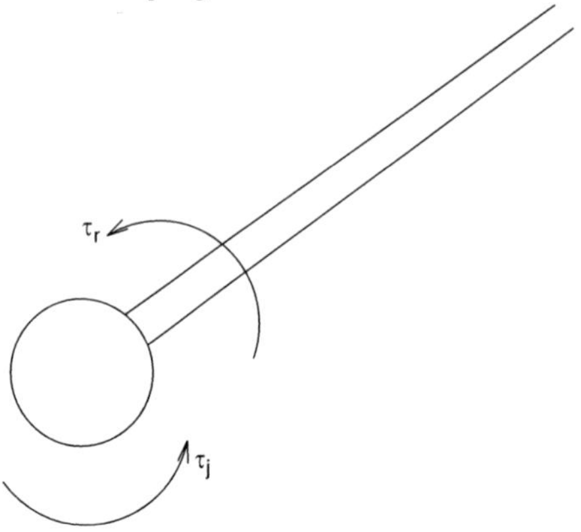 τ_r τ_j	No bending
Prismatic	Actuated spring 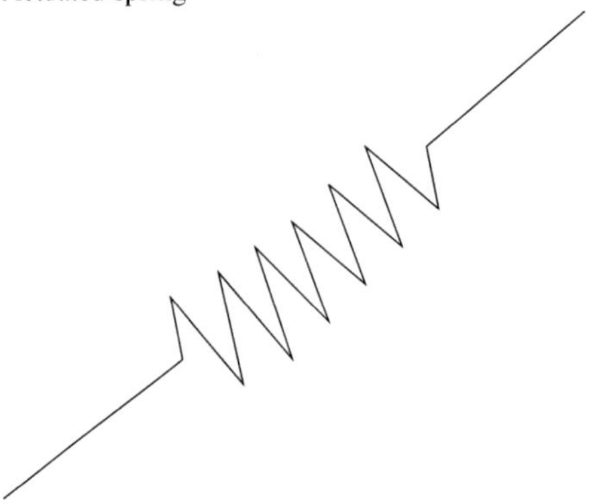	Uncertainty

will produce the same tip deflection as that of the beam under the load F. As it can be seen on the figure, the lumped model will lead to a different orientation of the tip of the beam. However, assuming that the deformation is small, angle θ will also be small, thus the difference in orientation between the original beam and the equivalent link can be neglected. Moreover, since in the mechanisms considered here, the legs are attached to the platform with spherical joints, there is not any moment presented at the spherical joint, hence, the end link orientation of the beam is irrelevant. Let δ be the deflection of the beam. Based on the Castiliano's theorem [143], one can build an equivalent rigid beam model based solely on the deflection of the free end. With a force F applied at the free end of the beam, the resulting deformation can be written as (see Fig. 4.1a)

$$\delta = \frac{FL^3}{3EI} \tag{4.10}$$

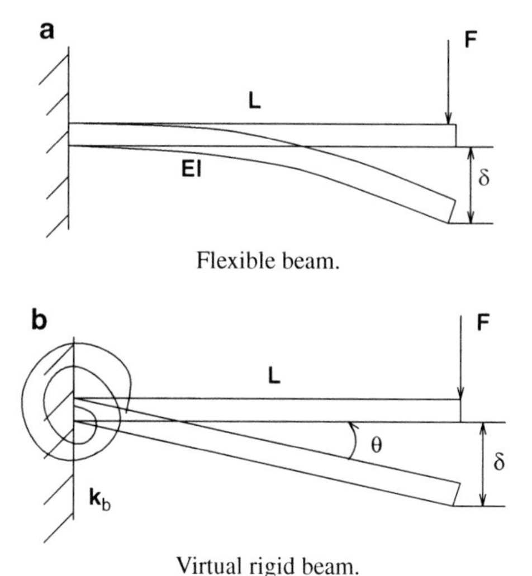

Fig. 4.1 Link deformation induced by wrench

and assuming small deformations, the corresponding rotational deformation of an equivalent rigid beam with a torsional spring would be

$$\theta \simeq \frac{\delta}{L}. \tag{4.11}$$

Let the deflection in both cases (Fig. 4.1a, 4.1b) be the same. Substituting (4.10) into (4.11), yields

$$\theta = \frac{FL^2}{3EI}, \tag{4.12}$$

where δ is the flexible beam's deflection at the free end and θ is the rigid beam's rotation around the joint.

Since the flexible beam model can be lumped into a torsional spring with equivalent stiffness k_b at the shoulder joint (Fig. 4.1b), based on the principle of work and energy, one has

$$\frac{1}{2}F\delta = \frac{1}{2}k_b(\theta)^2, \tag{4.13}$$

where k_b is the lumped stiffness of the flexible beam. Substituting (4.11) to (4.13), one obtains

$$FL\theta = k_b\theta^2 \tag{4.14}$$

or

$$k_b = \frac{FL}{\theta}. \tag{4.15}$$

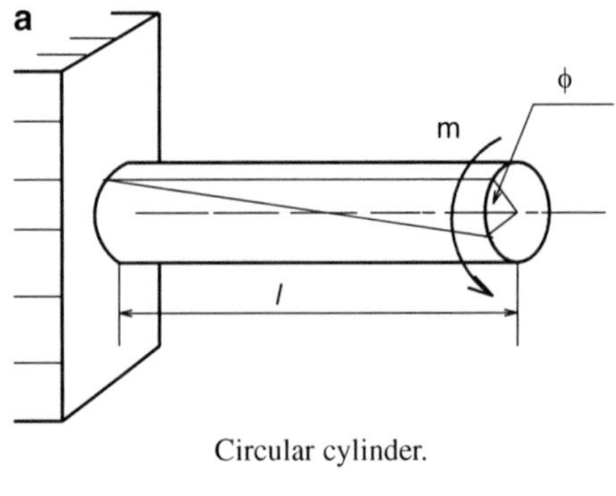

Circular cylinder.

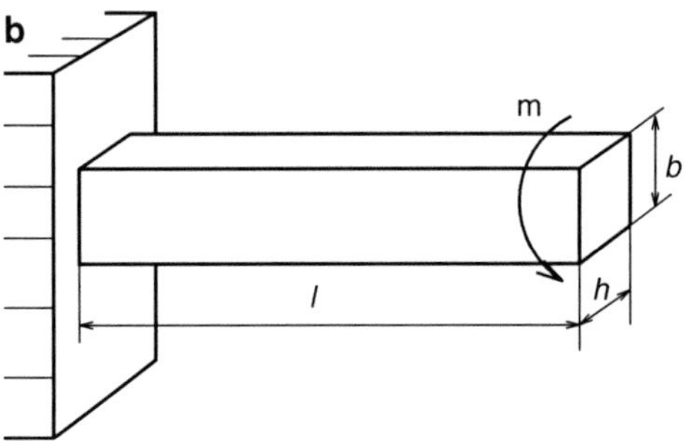

Rectangular parallelepiped.

Fig. 4.2 Link deformation induced by torque

Substituting (4.12) into (4.15), one obtains the equivalent stiffness for the flexible beam as

$$k_{\text{b}} = \frac{3EI}{L}. \tag{4.16}$$

Here the lumped stiffness expression for a single flexible beam undergoing twisting is addressed. A linear beam is shown in Fig. 4.2, where m(Nm) is the external torque, $G\,(\text{N/m}^2)$ the shear elastic modulus, l(m) the length of the beam, and $I\,(\text{m}^4)$ is the section moment of inertia of the beam. Similarly to the preceding section, the flexible beam is replaced by a rigid beam mounted at the end plus a torsional spring located at the end. The objective is to determine the equivalent torsional spring stiffness that will produce the same tip deflection as that of the beam under the load m. Assuming that the deformation is small, angle ϕ will also be small, then, with a twist m applied at the free end of the beam, the resulting deformation can be written as

$$\Delta\phi = \frac{ml}{GI} \quad \text{for circular cylinder,} \tag{4.17}$$

$$\Delta\phi = \frac{ml}{G\beta h^3 b} \quad \text{for rectangular parallelepiped,} \tag{4.18}$$

where b is the height of the flexible beam, h is the width of the flexible beam and β is a coefficient related to b and h. Since one has

$$m = k_t \Delta\phi \tag{4.19}$$

hence one can obtain the lumped stiffness k_t of the beam as

$$k_t = \frac{GI}{l} \quad \text{for circular cylinder,} \tag{4.20}$$

$$k_t = \frac{G\beta h^3 b}{l} \quad \text{for rectangular parallelepiped.} \tag{4.21}$$

4.2 Planar Two Degrees of Freedom Parallel Robotic Machine

As shown in Fig. 4.3, we take the case of revolute type into account. A planar two-degree-of-freedom mechanism can be used to position a point on the plane and the Cartesian coordinates associated with this mechanism are the position coordinates of one point of the platform, noted (x, y). Vector θ represents the actuated joint coordinates of the planar parallel mechanism and is defined as $\theta = [\theta_1, \theta_2, \ldots, \theta_n]^{\mathrm{T}}$, where n is the number of degrees of freedom of the mechanism studied, and the only actuated joints are those directly connected to the fixed link [59, 61, 133].

As illustrated in Fig. 4.3, a 2-dof planar parallel mechanism is constructed by four movable links and five revolute joints (noted as O_1 to O_5). The two links – whose length are l_1 and l_3 – are the input links. They are assumed to be flexible beams, and

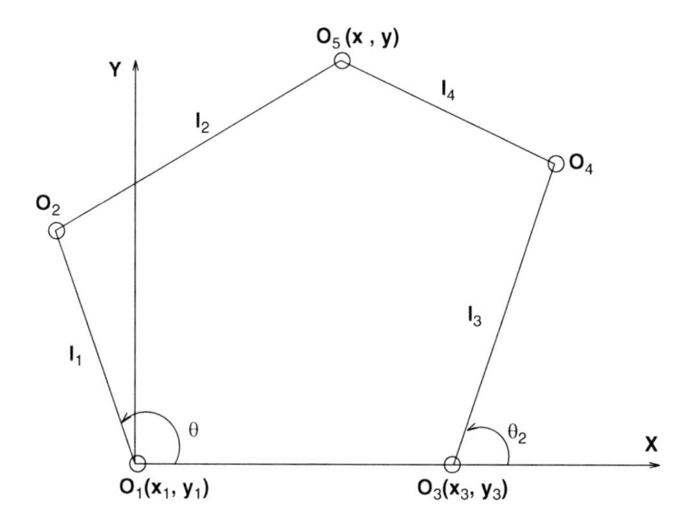

Fig. 4.3 A planar 2-dof parallel mechanism with revolute actuators

points O_1 and O_3 are the only actuated joints in this planar 2-dof parallel mechanism. The lengths of the other two links are denoted as l_2 and l_4, respectively. Point $O_5(x, y)$ is the point to be positioned by the mechanism. The origin of the fixed Cartesian coordinate system is located on joint O_1. (x_1, y_1) and (x_3, y_3) are the coordinates of points O_1 and O_3, respectively, and one has $x_1 = y_1 = y_3 = 0$.

At points O_2 and O_4, one has

$$x_2 = l_1 \cos\theta_1 + x_1, \tag{4.22}$$

$$y_2 = l_1 \sin\theta_1 + y_1, \tag{4.23}$$

$$x_4 = l_3 \cos\theta_2 + x_3, \tag{4.24}$$

$$y_4 = l_3 \sin\theta_2 + y_3. \tag{4.25}$$

From this figure, one obtains

$$l_2^2 = (x - x_2)^2 + (y - y_2)^2, \tag{4.26}$$

$$l_4^2 = (x - x_4)^2 + (y - y_4)^2. \tag{4.27}$$

Substituting (4.22) – (4.25) into (4.26) – (4.27), one gets

$$l_2^2 = (x - l_1\cos\theta_1)^2 + (y - l_1\sin\theta_1)^2, \tag{4.28}$$

$$l_4^2 = (x - (l_3\cos\theta_2 + x_3))^2 + (y - l_3\sin\theta_2)^2. \tag{4.29}$$

The kinematic relationship can be obtained as follows

$$\mathbf{F}(\theta, \mathbf{p}) = \begin{bmatrix} (x - l_1\cos\theta_1)^2 + (y - l_1\sin\theta_1)^2 - l_2^2 \\ (x - (l_3\cos\theta_2 + x_3))^2 + (y - l_3\sin\theta_2)^2 - l_4^2 \end{bmatrix} = 0. \tag{4.30}$$

Let

$$\dot{\theta} = \begin{bmatrix} \dot{\theta}_1 \\ \dot{\theta}_2 \end{bmatrix}, \quad \dot{\mathbf{p}} = \begin{bmatrix} \dot{x} \\ \dot{y} \end{bmatrix}. \tag{4.31}$$

One can obtain the Jacobian matrices of the parallel mechanism as

$$\mathbf{A} = \frac{\partial \mathbf{F}}{\partial \mathbf{p}}, \quad \mathbf{B} = \frac{\partial \mathbf{F}}{\partial \theta}. \tag{4.32}$$

In particular, the Jacobian matrices of this planar 2-dof parallel mechanism are as follows:

$$\mathbf{A} = \begin{bmatrix} (x - l_1\cos\theta_1) & (y - l_1\sin\theta_1) \\ (x - l_3\cos\theta_2 - x_3) & (y - l_3\sin\theta_2) \end{bmatrix}, \tag{4.33}$$

$$\mathbf{B} = \begin{bmatrix} (x\sin\theta_1 - y\cos\theta_1)l_1 & 0 \\ 0 & [(x - x_3)\sin\theta_2 - y\cos\theta_2]l_3 \end{bmatrix}. \tag{4.34}$$

The velocity equations can be written as $\mathbf{A}\dot{\mathbf{p}} + \mathbf{B}\dot{\theta} = 0$ and

$$\mathbf{J} = -\mathbf{B}^{-1}\mathbf{A} = \begin{bmatrix} a_1/d_1 & b_1/d_1 \\ a_2/d_2 & b_2/d_2 \end{bmatrix} \tag{4.35}$$

with

$$a_1 = x - l_1 \cos\theta_1, \tag{4.36}$$
$$a_2 = x - l_3 \cos\theta_2 - x_3, \tag{4.37}$$
$$b_1 = y - l_1 \sin\theta_1, \tag{4.38}$$
$$b_2 = y - l_3 \sin\theta_2, \tag{4.39}$$
$$d_1 = -(x \sin\theta_1 - y \cos\theta_1)l_1, \tag{4.40}$$
$$d_2 = -[(x - x_3)\sin\theta_2 - y\cos\theta_2]l_3. \tag{4.41}$$

In order to compute the Jacobian matrix of (4.35), one has to know the joint angles of Fig. 4.3 first. Therefore, it is necessary to calculate the inverse kinematics of this planar 2-dof parallel mechanism to determine the joint angles for any given end-effector position and orientation. Unlike many serial mechanisms, the calculation of the inverse kinematics of a parallel mechanism is generally straightforward.

From (4.28), one obtains

$$2l_1 x \cos\theta_1 + 2l_1 y \sin\theta_1 = x^2 + y^2 + L_1^2 - L_2^2, \tag{4.42}$$

therefore, one can obtain θ_1 as follow

$$\sin\theta_1 = \frac{BC + K_1 A\sqrt{A^2 + B^2 - C^2}}{A^2 + B^2}, \tag{4.43}$$

$$\cos\theta_1 = \frac{AC - K_1 B\sqrt{A^2 + B^2 - C^2}}{A^2 + B^2}, \tag{4.44}$$

where

$$A = 2l_1 x, \tag{4.45}$$
$$B = 2l_1 y, \tag{4.46}$$
$$C = x^2 + y^2 + L_1^2 - L_2^2, \tag{4.47}$$
$$K_1 = \pm 1 \tag{4.48}$$

and K_1 is the branch index, which can be used to distinguish the four branches of the inverse kinematic problem. In the same way, from (4.29), one obtains

$$2l_3(x - x_3)\cos\theta_2 + 2l_3 y \sin\theta_2 = (x - x_3)^2 + y^2 + l_3^2 - l_4^2. \tag{4.49}$$

Hence one obtains the joint angle θ_2 as

$$\sin\theta_2 = \frac{BC + K_2 A\sqrt{A^2 + B^2 - C^2}}{A^2 + B^2}, \tag{4.50}$$

$$\cos\theta_2 = \frac{AC - K_2 B\sqrt{A^2 + B^2 - C^2}}{A^2 + B^2}, \tag{4.51}$$

where

$$A = 2l_3(x - x_3), \tag{4.52}$$

$$B = 2l_3 y, \tag{4.53}$$

$$C = (x - x_3)^2 + y^2 + l_3^2 - l_4^2, \tag{4.54}$$

$$K_2 = \pm 1. \tag{4.55}$$

Again, K_2 is the branch index.

Assume the actuator stiffnesses of O_1 and O_3 are k_1 and k_1', respectively, and the lumped stiffness for beam O_1O_2 and O_3O_4 are k_b and k_b'. Then the compound stiffness at points O_1 and O_3 are written as

$$k = \frac{k_1 k_\mathrm{b}}{k_1 + k_\mathrm{b}}, \tag{4.56}$$

$$k' = \frac{k_1' k_\mathrm{b}'}{k_1' + k_\mathrm{b}'}, \tag{4.57}$$

where k, k' are the total stiffnesses at the active joint, k_1, k_1' are the actuator stiffnesses and k_b, k_b' are the lumped stiffnesses as indicated in (4.16). One can find the kinetostatic model for this planar 2-dof parallel mechanism by using (4.7), i.e.,

$$\mathbf{K}_\mathrm{C} = \mathbf{J}^\mathrm{T}\mathbf{K}_\mathrm{J}\mathbf{J}, \tag{4.58}$$

where $\mathbf{K_J}$ is the joint stiffness matrix of the parallel mechanism and $\mathbf{J}$ is the Jacobian matrix of this planar 2-dof parallel mechanism.

The analysis described above is now used to obtain the stiffness maps for this planar 2-dof parallel mechanism. The maps are drawn on a section of the workspace of the variation of the end-effector's position.

A program has been written with the software Matlab. Given the values of $l_1 = l_4 = 0.5\,\mathrm{m}$, $l_2 = 0.6\,\mathrm{m}$, $l_3 = 0.8\,\mathrm{m}$, and $O_1O_3 = 0.7\,\mathrm{m}$. The contour graph can be shown in Fig. 4.4.

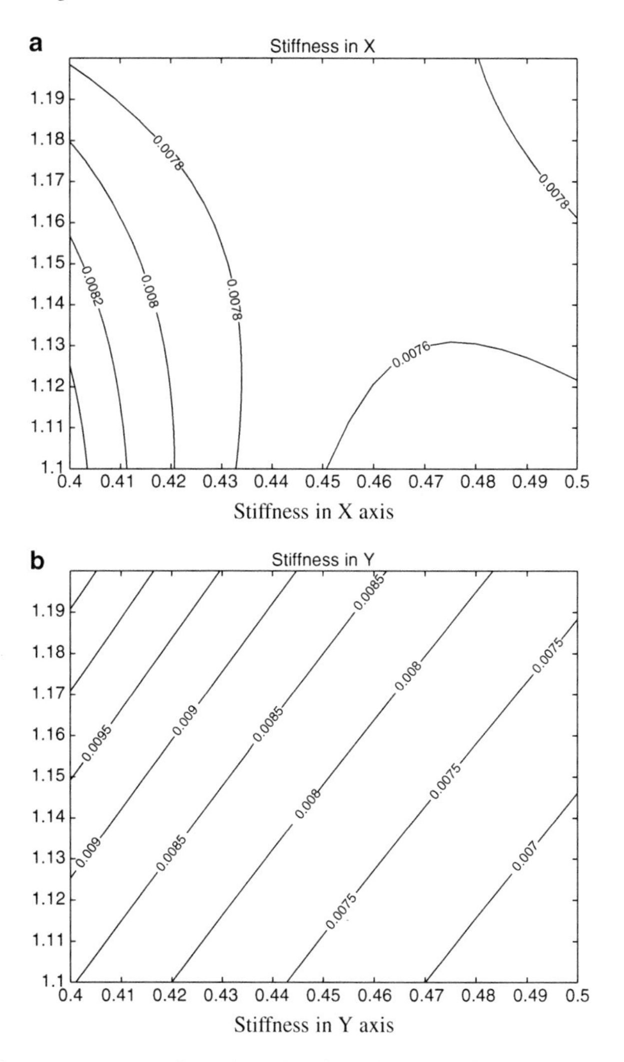

Fig. 4.4 Stiffness contour graph for a planar 2-dof parallel mechanism with revolute actuators

4.3 Planar Three Degrees of Freedom Parallel Robotic Machine

A symmetric mechanism identical to the one studied in [56] and [58] is now analyzed with the procedure described above. The characteristics of this mechanism are as follows: Points A_i, $i = 1, 2, 3$ and points B_i, $i = 1, 2, 3$ (Fig. 4.5) are, respectively, located on the vertices of an equilateral triangle and that the minimum and maximum lengths of each of the legs are the same. The mechanism is therefore completely symmetric. The dimensions and the stiffness of each leg are given in Table 4.2.

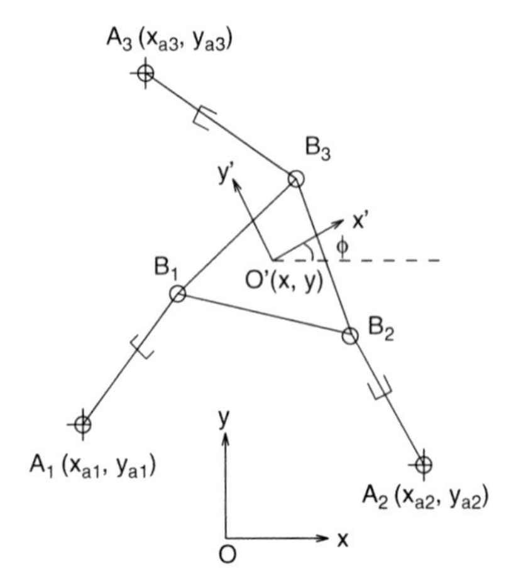

Fig. 4.5 A planar 3-dof parallel mechanism with prismatic actuators

Table 4.2 Geometric properties of symmetric planar parallel mechanism (all length units in mm and stiffness units in N/m)

i	x_{ai}	y_{ai}	x_{bi}	y_{bi}	k_i
1	$-1/2$	$-\sqrt{3}/6$	$-1/12$	$-\sqrt{3}/36$	1,000
2	$1/2$	$-\sqrt{3}/6$	$1/12$	$-\sqrt{3}/36$	1,500
3	0	$\sqrt{3}/3$	0	$\sqrt{3}/18$	700

Since one has

$$x_i = x - L\cos\phi_i - x_{ai}, \quad i = 1,2,3, \tag{4.59}$$

$$y_i = y - L\sin\phi_i - y_{ai}, \quad i = 1,2,3, \tag{4.60}$$

$$p_i = \sqrt{x_i^2 + y_i^2}, \quad i = 1,2,3, \tag{4.61}$$

where L is the length of the gripper and p_i is the length of the leg. The Jacobian matrix is given by [56] as follows

$$\mathbf{J} = \begin{bmatrix} a_1/p_1 & b_1/p_1 & c_1/p_1 \\ a_2/p_2 & b_2/p_2 & c_2/p_2 \\ a_3/p_3 & b_3/p_3 & c_3/p_3 \end{bmatrix} \tag{4.62}$$

with

$$a_i = x - x_{ai} - L\cos\phi_i, \tag{4.63}$$

$$b_i = y - y_{ai} - L\sin\phi_i, \tag{4.64}$$

$$c_i = (x - x_{ai})L\sin\phi_i - (y - y_{ai})L\cos\phi_i. \tag{4.65}$$

Hence, according to (4.7), one can find the stiffness model for this planar three-degrees-of-freedom parallel mechanism.

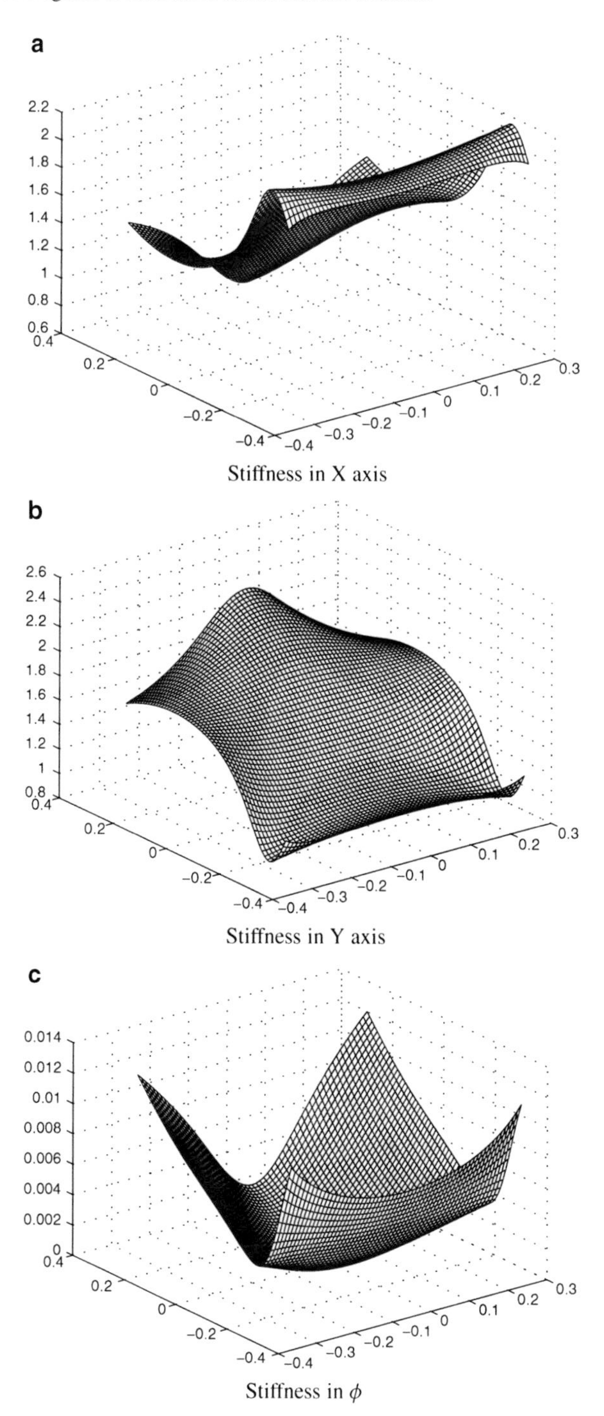

Fig. 4.6 Stiffness mesh graphs for a planar 3-dof parallel mechanism with prismatic actuators ($\phi = 0$)

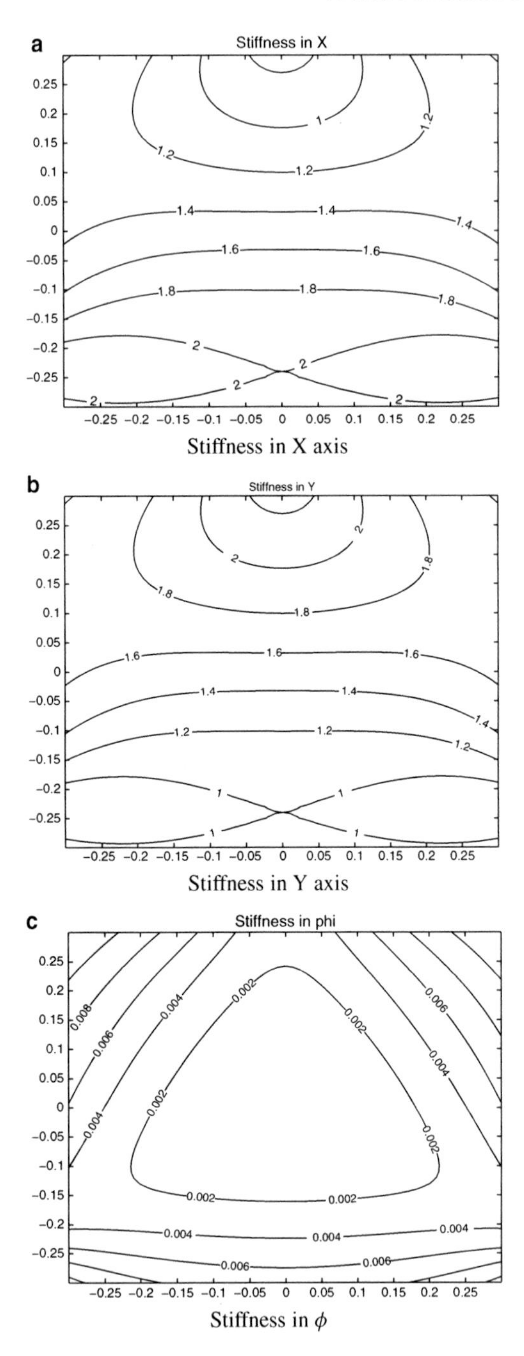

Fig. 4.7 Stiffness contour graphs for a planar 3-dof parallel mechanism with prismatic actuators ($\phi = 0$)

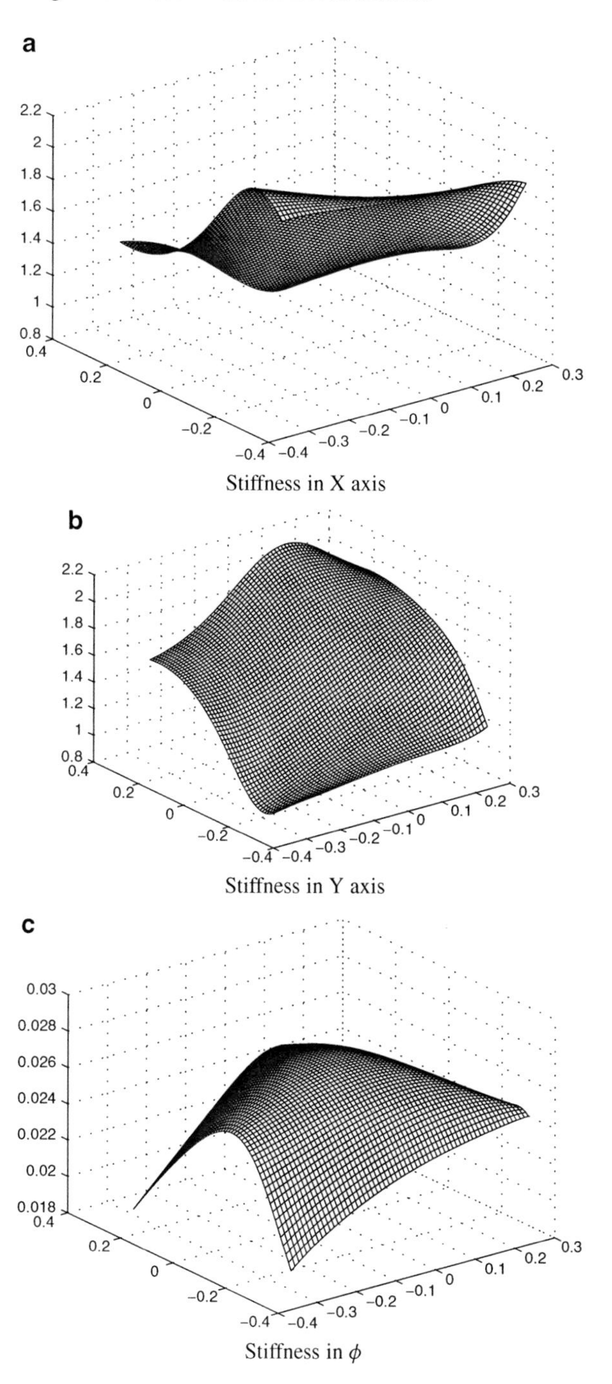

Fig. 4.8 Stiffness mesh graphs for a planar 3-dof parallel mechanism with prismatic actuators ($\phi = \pi/2$)

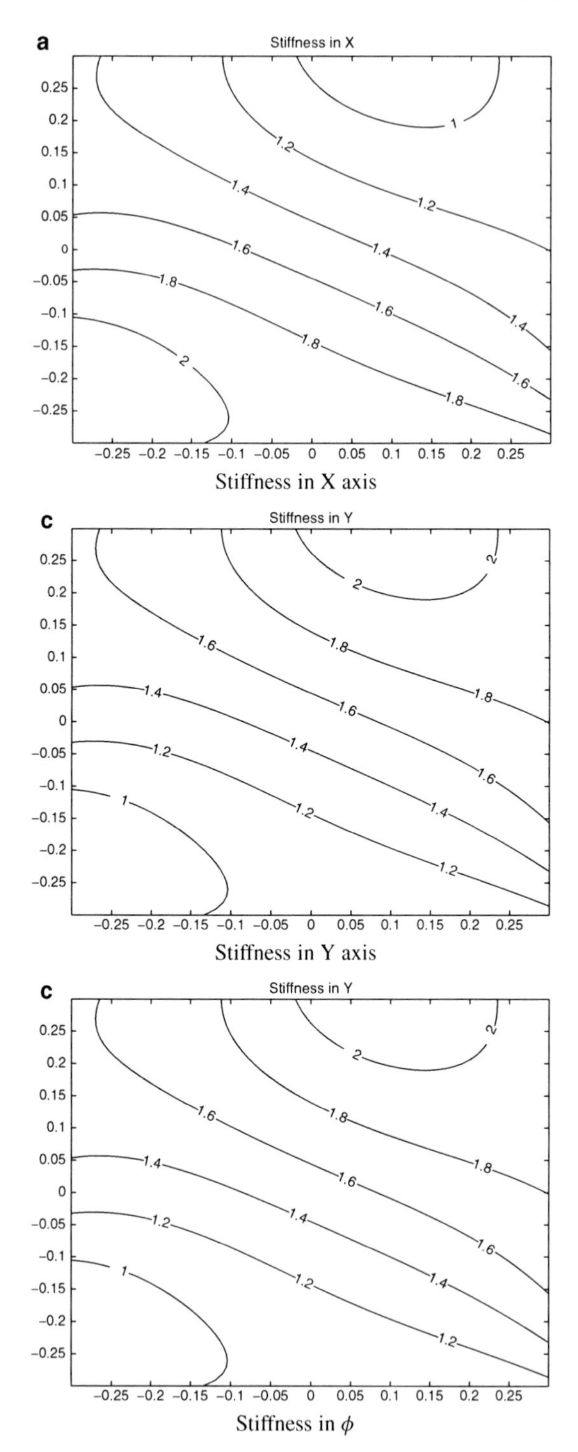

Fig. 4.9 Stiffness contour graphs for a planar 3-dof parallel mechanism with prismatic actuators ($\phi = \pi/2$)

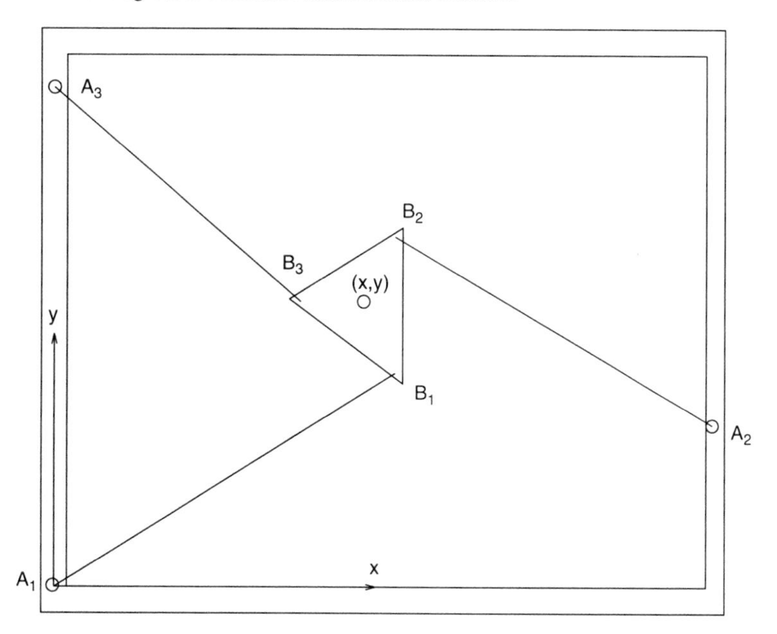

Fig. 4.10 Validation model of the planar 3-dof parallel mechanism in Pro/Motion

Table 4.3 Geometric properties of planar parallel mechanism (all units in mm)

i	x_{ai}	y_{ai}	x_{bi}	y_{bi}	k_i
1	0	0	84.547	48.464	400
2	150	49	84.547	81.536	400
3	0	130	55.91	65	400

The above model is now used to obtain the stiffness maps for this planar 3-dof parallel mechanism. Given the values shown in Table 4.2, one can obtain the stiffness contour and mesh graphs in x, y, and ϕ shown in Figs. 4.6 – 4.9.

One can find from the stiffness map that the symmetric mechanism is in a singular configuration when positioned at the center of the workspace. Also, from such stiffness maps, one can determine which regions of the workspace will satisfy some stiffness criteria. From the mesh graphs, one can view the stiffness distribution more intuitively.

A model (Fig. 4.10) for this planar 3-dof parallel mechanism has been built using the software Pro/Engineer to simulate the physical structure on Pro/Motion.

With the geometric properties given in Table 4.3 and the center of the triangle located at $(75, 65)$, after applying the forces and torque $F_x = 100$ N, $F_y = 100$ N, $\tau = 60$ Nm at the center of the triangle, the three legs deform. One obtains the deformation of the center using Pro/Motion as $\Delta x = 0.09697$ mm, $\Delta y = 0.14959$ mm, $\Delta\phi = -0.0020$. Meanwhile, the results obtained from the equations developed in the previous section are $\Delta x = 0.0962$ mm, $\Delta y = 0.1548$ mm,

$\Delta\phi = -0.0020$. This shows that the results from Pro/Motion and the kinetostatic model are very close to each other.

4.4 Conclusions

A general stiffness model for fully- parallel mechanisms with different actuator stiffnesses has been presented in this chapter. It has been shown that this general stiffness model can be used to evaluate the stiffness properties of parallel mechanisms. Examples have been given to illustrate how this model is used. Meanwhile, the lumped models for joints and links are proposed. They can be applied to establish kinetostatic models for both 2-dof and 3-dof mechanisms which are also mentioned in this chapter. Finally, the reliability of the stiffness model has been demonstrated using the computer program Pro/Engineer.

Chapter 5
Spatial Parallel Robotic Machines with Prismatic Actuators

5.1 Preamble

In this chapter, we first introduce a fully six degrees of freedom fully-parallel robotic machine with prismatic actuators. Then several new types of parallel mechanisms with prismatic actuators whose degree of freedom is dependent on a constraining passive leg connecting the base and the platform is analyzed. The mechanisms are a series of n-dof parallel mechanisms which consist of n identical actuated legs with six degrees of freedom and one passive leg with n degrees of freedom connecting the platform and the base. This series of mechanisms has the characteristics of reproduction since they have identical actuated legs, thus, the entire mechanism essentially consists of repeated parts, offering price benefits for manufacturing, assembling, and maintenance.

A simple method for the stiffness analysis of spatial parallel mechanisms is presented using a lumped parameter model. Although it is essentially general, the method is specifically applied to spatial parallel mechanisms. A general kinematic model is established for the analysis of the structural rigidity and accuracy of this family of mechanisms. One can improve the rigidity of this type of mechanism through optimization of the link rigidities and geometric dimensions to reach the maximized global stiffness and precision. In what follows, the geometric model of this class of mechanisms is first introduced. The virtual joint concepts are employed to account for the compliance of the links. A general kinematic model of the family of parallel mechanisms is then established and analyzed using the lumped-parameter model. Equations allowing the computation of the equivalent joint stiffnesses are developed. Additionally, the inverse kinematics and velocity equations are given for both rigid-link and flexible-link mechanisms. Finally, examples for 3-dof, 4-dof, 5-dof, and 6-dof are given in detail to illustrate the results.

D. Zhang, *Parallel Robotic Machine Tools*, DOI 10.1007/978-1-4419-1117-9_5,

5.2 Six Degrees of Freedom Parallel Robotic Machine with Prismatic Actuators

5.2.1 Geometric Modeling and Inverse Kinematics

A 6-dof parallel mechanism and its joint distributions both on the base and on the platform are shown in Figs. 5.1–5.3. This mechanism consists of six identical variable length links, connecting the fixed base to a moving platform. The kinematic chains associated with the six legs, from base to platform, consist of a fixed Hooke joint, a moving link, an actuated prismatic joint, a second moving link, and a spherical joint attached to the platform. It is also assumed that the vertices on the base and on the platform are located on circles of radii R_b and R_p, respectively.

A fixed reference frame $O - xyz$ is connected to the base of the mechanism and a moving coordinate frame $P - x'y'z'$ is connected to the platform. In Fig. 5.2, the points of attachment of the actuated legs to the base are represented with B_i and the points of attachment of all legs to the platform are represented with P_i, with $i = 1, \ldots, 6$, while point P is located at the center of the platform with the coordinate of $P(x, y, z)$.

The Cartesian coordinates of the platform are given by the position of point P with respect to the fixed frame, and the orientation of the platform (orientation of frame $P - x'y'z'$ with respect to the fixed frame), represented by three Euler angles ϕ, θ, and ψ or by the rotation matrix $\mathbf{Q}$.

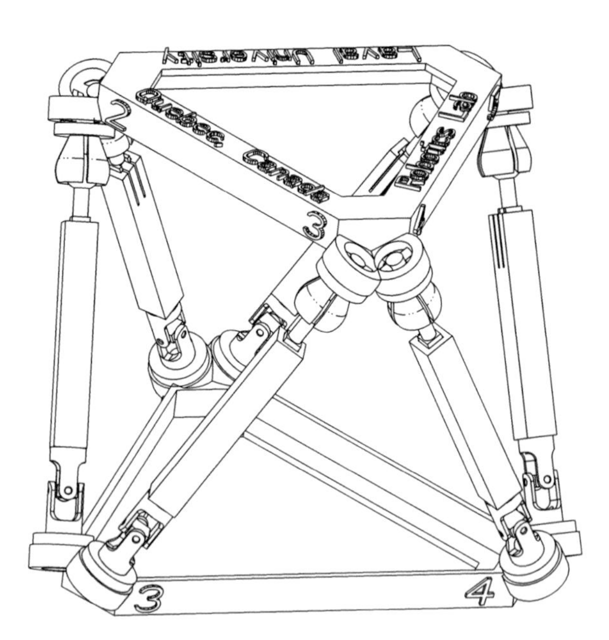

Fig. 5.1 CAD model of the spatial 6-dof parallel mechanism with prismatic actuators (Figure by Thierry Laliberté and Gabriel Coté)

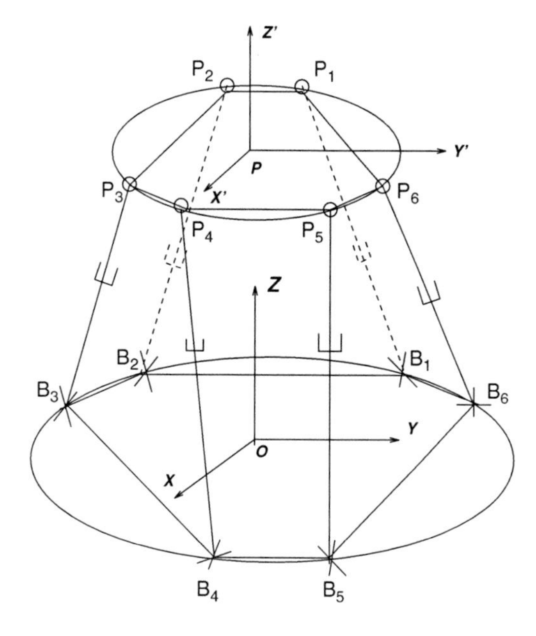

Fig. 5.2 Schematic representation of the spatial 6-dof parallel mechanism with prismatic actuators

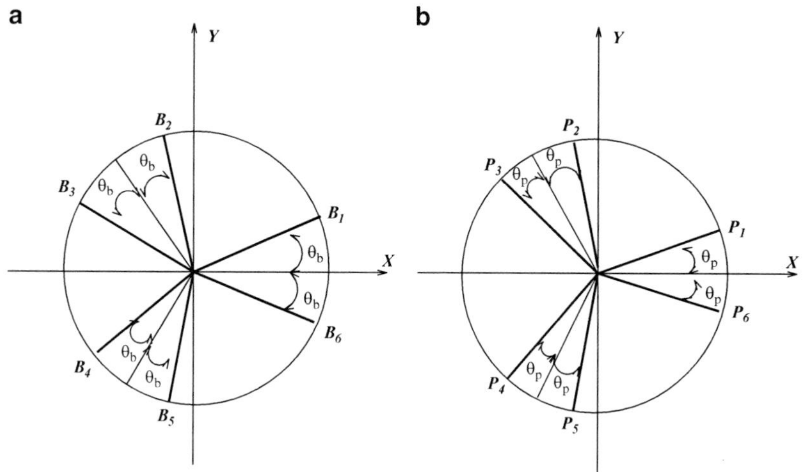

Fig. 5.3 Position of the attachment points: (**a**) on the base, (**b**) on the platform

If the coordinates of point B_i in the fixed frame are represented by vector $\mathbf{b}_i$, then we have

$$\mathbf{p}_i = \begin{bmatrix} x_i \\ y_i \\ z_i \end{bmatrix}, \; \mathbf{r}_i' = \begin{bmatrix} R_{\mathrm{p}} \cos \theta_{\mathrm{p}i} \\ R_{\mathrm{p}} \sin \theta_{\mathrm{p}i} \\ 0 \end{bmatrix}, \; \mathbf{p} = \begin{bmatrix} x \\ y \\ z \end{bmatrix}, \; \mathbf{b}_i = \begin{bmatrix} R_{\mathrm{b}} \cos \theta_{\mathrm{b}i} \\ R_{\mathrm{b}} \sin \theta_{\mathrm{b}i} \\ 0 \end{bmatrix}, \tag{5.1}$$

where $\mathbf{p}_i$ is the position vector of point P_i expressed in the fixed coordinate frame whose coordinates are defined as (x_i, y_i, z_i), $\mathbf{r}_i^{'}$ is the position vector of point P_i expressed in the moving coordinate frame, and $\mathbf{p}$ is the position vector of point P expressed in the fixed frame as defined above, and

$$\theta_{\mathrm{b}i} = \begin{bmatrix} \theta_{\mathrm{b1}} \\ \theta_{\mathrm{b2}} \\ \theta_{\mathrm{b3}} \\ \theta_{\mathrm{b4}} \\ \theta_{\mathrm{b5}} \\ \theta_{\mathrm{b6}} \end{bmatrix} = \begin{bmatrix} \theta_{\mathrm{b}} \\ 2\pi/3 - \theta_{\mathrm{b}} \\ 2\pi/3 + \theta_{\mathrm{b}} \\ 4\pi/3 - \theta_{\mathrm{b}} \\ 4\pi/3 + \theta_{\mathrm{b}} \\ -\theta_{\mathrm{b}} \end{bmatrix}, \ \theta_{\mathrm{p}i} = \begin{bmatrix} \theta_{\mathrm{p1}} \\ \theta_{\mathrm{p2}} \\ \theta_{\mathrm{p3}} \\ \theta_{\mathrm{p4}} \\ \theta_{\mathrm{p5}} \\ \theta_{\mathrm{p6}} \end{bmatrix} = \begin{bmatrix} \theta_{\mathrm{p}} \\ 2\pi/3 - \theta_{\mathrm{p}} \\ 2\pi/3 + \theta_{\mathrm{p}} \\ 4\pi/3 - \theta_{\mathrm{p}} \\ 4\pi/3 + \theta_{\mathrm{p}} \\ -\theta_{\mathrm{p}} \end{bmatrix}. \tag{5.2}$$

Similarly, the solution of the inverse kinematic of this mechanism can be written as

$$\rho_i^2 = (\mathbf{p}_i - \mathbf{b}_i)^{\mathrm{T}}(\mathbf{p}_i - \mathbf{b}_i), \quad i = 1, \ldots, 6. \tag{5.3}$$

5.2.2 Global Velocity Equation

Since the mechanism is actuated in parallel, one has the velocity equation as

$$\mathbf{At} = \mathbf{B}\dot{\rho}, \tag{5.4}$$

where vectors $\dot{\rho}$ and $\mathbf{t}$ are defined as

$$\dot{\rho} = \left[\dot{\rho}_1 \cdots \dot{\rho}_6 \right]^{\mathrm{T}}, \tag{5.5}$$

$$\mathbf{t} = \left[\omega^{\mathrm{T}} \ \dot{\mathbf{p}}^{\mathrm{T}} \right]^{\mathrm{T}}, \tag{5.6}$$

where ω and $\dot{\mathbf{p}}$ are the angular velocity and velocity of one point of the platform, respectively, and

$$\mathbf{A} = \left[\mathbf{m}_1 \ \mathbf{m}_2 \ \mathbf{m}_3 \ \mathbf{m}_4 \ \mathbf{m}_5 \ \mathbf{m}_6 \right]^{\mathrm{T}} \tag{5.7}$$

$$\mathbf{B} = \mathrm{diag}[\rho_1, \rho_2, \rho_3, \rho_4, \rho_5, \rho_6] \tag{5.8}$$

and $\mathbf{m}_i$ is a six-dimensional vector expressed as

$$\mathbf{m}_i = \begin{bmatrix} (\mathbf{Q}\mathbf{r'}_i) \times (\mathbf{p}_i - \mathbf{b}_i) \\ (\mathbf{p}_i - \mathbf{b}_i) \end{bmatrix}. \tag{5.9}$$

Therefore, Jacobian matrix $\mathbf{J}$ can be written as

$$\mathbf{J} = \mathbf{B}^{-1}\mathbf{A}. \tag{5.10}$$

The derivation of the relationship between Cartesian velocities and joint rates is thereby completed.

5.2.3 *Stiffness Model*

Since the mechanism is fully parallel, the stiffness of the mechanism has been obtained as

$$\mathbf{K} = \mathbf{J}^{\mathrm{T}}\mathbf{K}_{\mathrm{J}}\mathbf{J}. \tag{5.11}$$

Given the geometric properties as

$$\begin{aligned}
\theta_{\mathrm{p}} &= 22.34°, \theta_{\mathrm{b}} = 42.883°, \\
R_{\mathrm{p}} &= 6\,\mathrm{cm}, R_{\mathrm{b}} = 15\,\mathrm{cm}, \\
K_i &= -1, \quad i = 1, \ldots, 6 \\
k_{i1} &= 1,000\,\mathrm{N/m}, \quad i = 1, \ldots, 6\ ,
\end{aligned}$$

where k_{i1} is the actuator stiffness, and the Cartesian coordinates are given by

$$\begin{aligned}
x \in [-4, 4]\,\mathrm{cm}, y \in [-4, 4]\,\mathrm{cm}, z = 51\,\mathrm{cm}, \\
\phi = 0, \theta = 0, \psi = 0,
\end{aligned}$$

The stiffness model described above is now used to obtain the stiffness mappings. Fig. 5.4 shows the stiffness mappings on a section of the workspace of the platform. From such plots one can determine which regions of the workspace will satisfy some stiffness criteria.

Another example of mechanism is the INRIA "left-hand" prototype described in [57]. The dimensions are given in Table 5.1, the stiffness mappings and mesh graphs are illustrated in Fig. 5.5, the results show the same trends of stiffness as in [57].

From the graphs, one observes that K_{θ_x} and K_{θ_y}, K_x and K_y are symmetric with respect to each other, and in Fig. 5.4a the stiffness in X becomes higher when the platform moves further from the Y-axis. This was to be expected because when the platform moves aside along the X-axis, the projection of the legs on this axis becomes larger, and the mechanism is stiffer in Y. The same reasoning applies to Fig. 5.4b for the stiffness in Y.

In Fig. 5.4d, 5.4e, the torsional stiffnesses in θ_x and θ_y are shown, the stiffness is larger when it moves further from the Y-axis. In Fig. 5.4c, the stiffness in Z is higher near the center of the workspace, which is the best position for supporting vertical loads. It can also be noted that the stiffness in Z is much larger than the

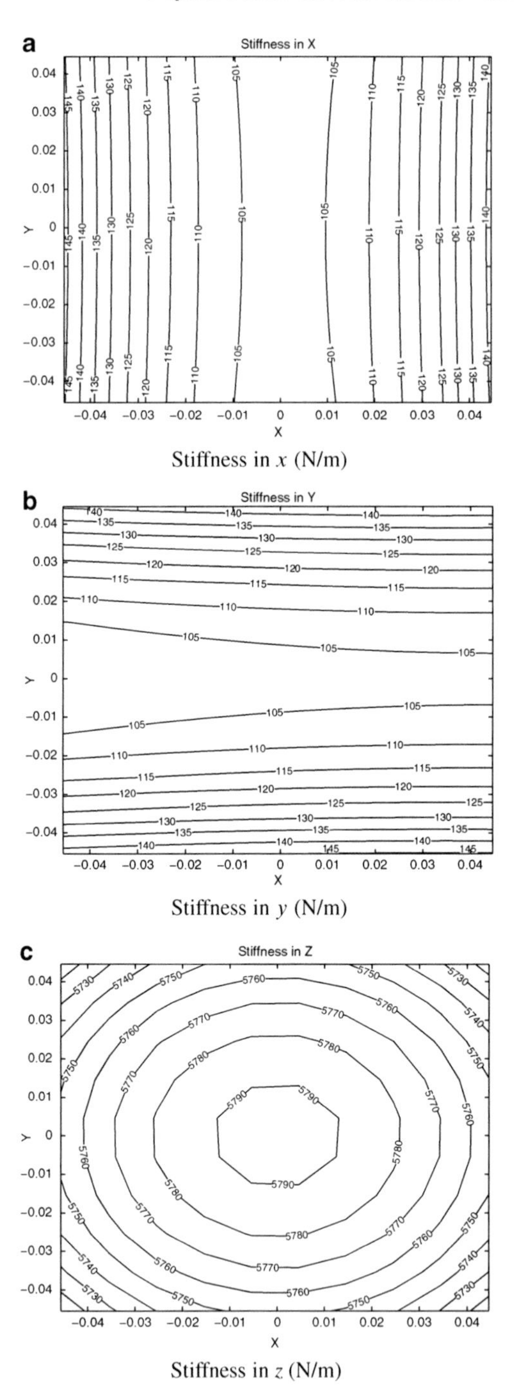

Fig. 5.4 Stiffness mappings of the spatial 6-dof parallel mechanism with prismatic actuators (all length units in m)

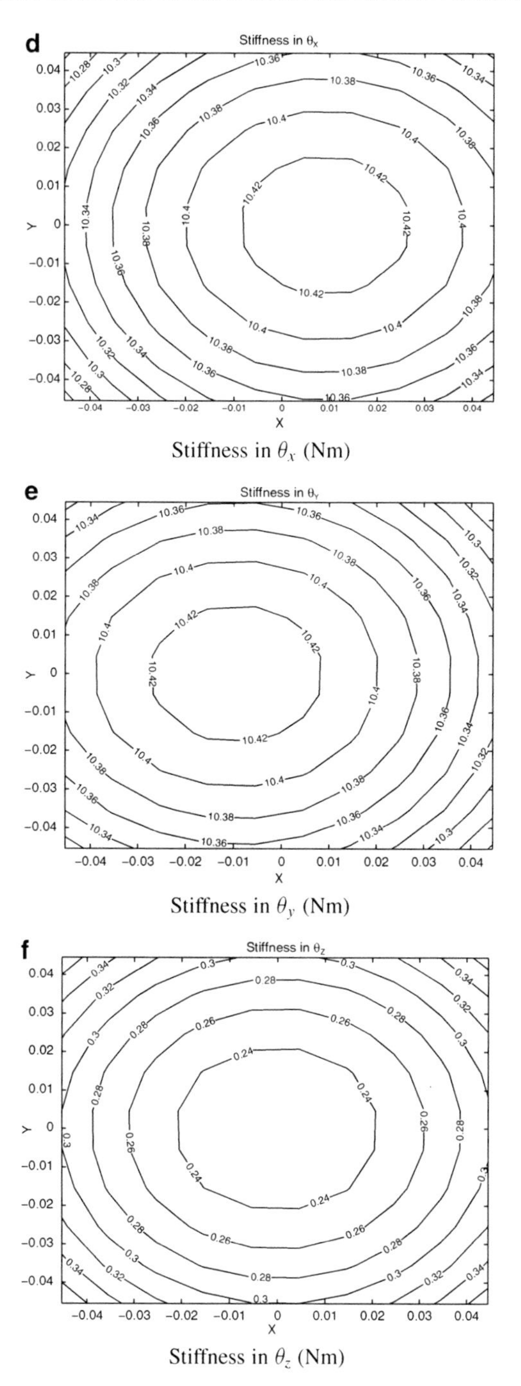

Stiffness in θ_x (Nm)

Stiffness in θ_y (Nm)

Stiffness in θ_z (Nm)

Fig. 5.4 (continued)

Table 5.1 Geometric properties of the INRIA prototype (all lengths are in centimeters)

i	1	2	3	4	5	6
b_{ix}	9.258	13.258	4.000	−4.000	−13.258	−9.258
b_{iy}	9.964	3.036	−13.000	−13.000	3.036	9.964
b_{iz}	2.310	2.310	2.310	2.310	2.310	2.310
x_i	3.000	7.822	4.822	−4.822	−7.822	−3.000
y_i	7.300	−1.052	−6.248	−6.248	−1.052	7.300
z_i	−3.710	−3.710	−3.710	−3.710	−3.710	−3.710
L_i	51	51	51	51	51	51

stiffness in the X or Y directions. This is due to the architecture chosen, which aims at supporting heavy objects in an environment where the gravity is acting along the negative direction of Z-axis. All these are in accordance with what would be intuitively expected.

Table 5.2 shows the variation of the stiffness with K_{actuator}. Clearly, the Cartesian stiffnesses in each direction are increased with the improvement of the actuator stiffness.

5.3 General Kinematic Model of *n* Degrees of Freedom Parallel Mechanisms with a Passive Constraining Leg and Prismatic Actuators

5.3.1 Geometric Modeling and Lumped Compliance Model

5.3.1.1 Geometric Modeling

An example of parallel mechanisms belonging to the family of mechanisms studied in this chapter is shown in Figs. 5.6 and 5.7. It is a 5-dof parallel mechanism with prismatic actuators. This mechanism consists of six kinematic chains, including five variable length legs with identical topology and one passive leg which connects the fixed base to the moving platform. In this 5-dof parallel mechanism, the kinematic chains associated with the five identical legs consist, from base to platform, of a fixed Hooke joint, a moving link, an actuated prismatic joint, a second moving link, and a spherical joint attached to the platform. The sixth chain (central leg) connecting the base center to the platform is a passive constraining leg and has an architecture different from the other chains. It consists of a revolute joint attached to the base, a moving link, a Hooke joint, a second moving link, and another Hooke joint attached to the platform. This last leg is used to constrain the motion of the platform to only 5 dof. This mechanism could be built using only five legs, i.e., by removing one of the five identical legs and actuating the first joint of the passive constraining leg. However, the uniformity of the actuation would be lost.

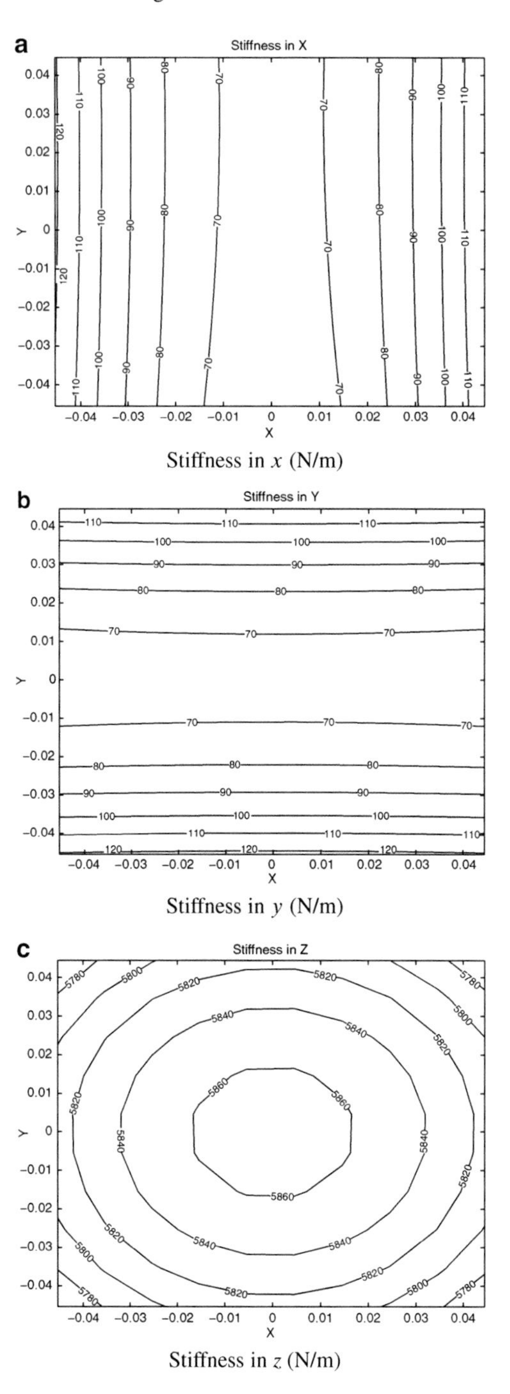

Fig. 5.5 Stiffness mappings of the spatial 6-dof parallel mechanism with prismatic actuators (using data of INRIA prototype) (all length units in m)

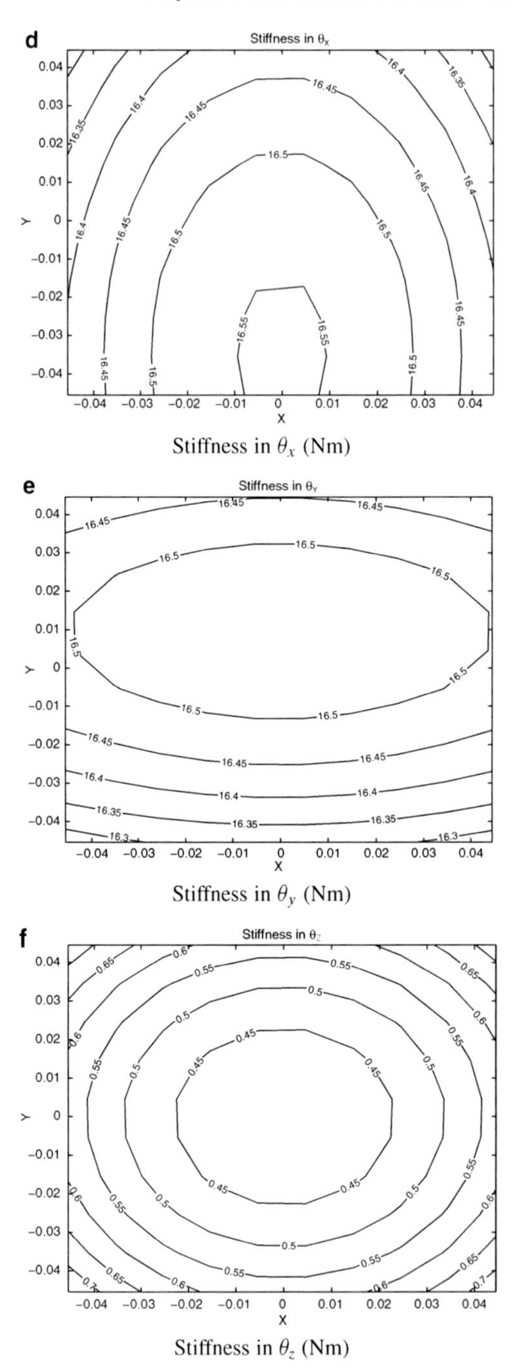

Fig. 5.5 (continued)

Table 5.2 The Cartesian stiffness as a function of the actuator stiffness

K_{actuator}	K_x	K_y	K_z	K_{θ_x}	K_{θ_y}	K_{θ_z}
200	20.5936	20.5936	1158.81	2.08586	2.08586	0.0444375
600	61.7809	61.7809	3476.44	6.25759	6.25759	0.133313
1,000	102.968	102.968	5794.06	10.4293	10.4293	0.222188
2,000	205.936	205.936	11588.1	20.8586	20.8586	0.444375
3,000	308.904	308.904	17382.2	31.2879	31.2879	0.666563
4,000	411.872	411.872	23176.3	41.7173	41.7173	0.888751
6,000	617.809	617.809	34764.4	62.5759	62.5759	1.33313

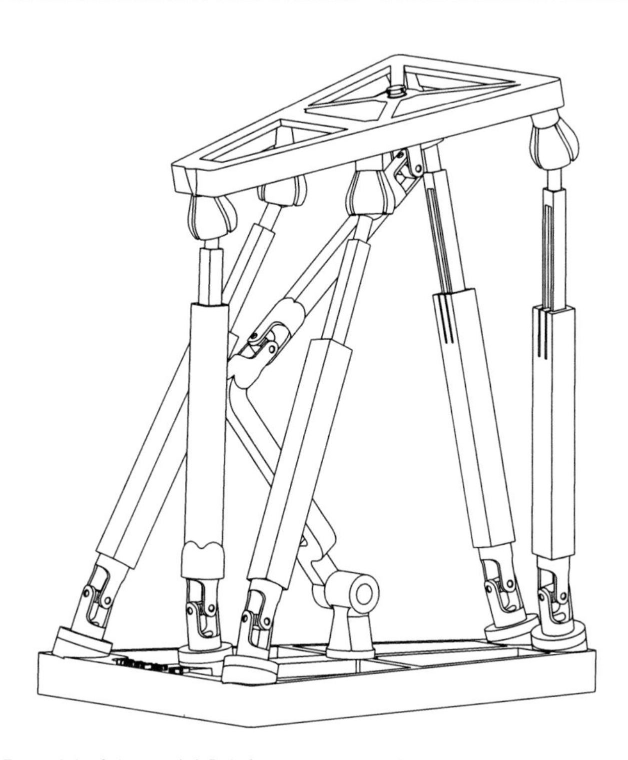

Fig. 5.6 CAD model of the spatial 5-dof parallel mechanism with prismatic actuators (Figure by Gabriel Coté)

Similarly, families of 3-dof and 4-dof parallel mechanisms can be built using three or four identical legs with 6 dof and one passive constraining leg with 3 or 4 dof, respectively, and they will also be discussed in this chapter. The aim of using the passive leg is to limit the degrees of freedom to the desired ones. Since the external loads on the platform will induce bending and/or torsion in the passive leg, its mechanical design is a very important issue which can be addressed using the kinetostatic model proposed here. It should be noted, however, that the final geometry and mechanical design of the passive leg may be significantly different from the generic representation given in Fig. 5.6.

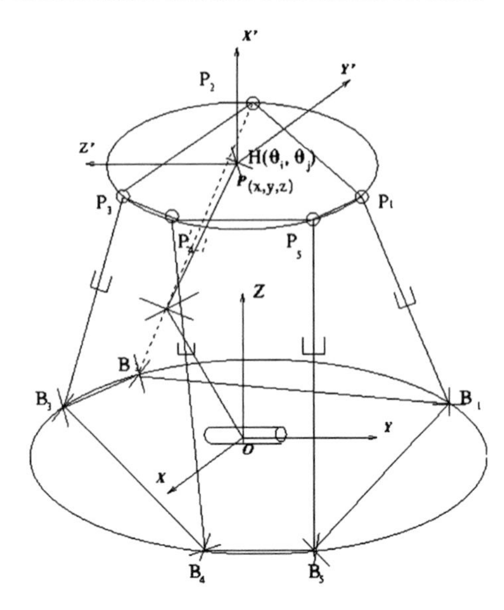

Fig. 5.7 Schematic representation of the spatial 5-dof parallel mechanism with prismatic actuators

Table 5.3 Lumped joint compliance models for spatial system

Joint type	If actuated, the equivalent model	If unactuated, the equivalent model
Spherical	N/A	No transformation
Hooke	N/A	No transformation
Revolute	4 Torsional springs	2 torsional spring
	4 Torsional springs	2 Torsional springs

5.3.1.2 Lumped Models for Joint and Link Compliances

In order to obtain a simple kinetostatic model, link compliances are lumped at the joints as described in [15]. In this framework, link bending stiffnesses are replaced by equivalent torsional springs located at virtual joints (joints with dotted lines in Table 5.3), as illustrated in Table 5.3. Actuator stiffnesses are also included and modeled as torsional or linear springs for revolute and prismatic actuators, respectively. For instance, the actuated revolute joint with flexible link can be lumped to 4 (four) torsional springs including 2 (two) bendings, 1 (one) twist, and 1 (one) actuator. The details are described in [15].

5.3.2 *Inverse Kinematics*

Since the platform of the mechanism has n degrees of freedom, only n of the six Cartesian coordinates of the platform are independent. For the 5-dof mechanism of Fig. 5.6, the independent coordinates have been chosen for convenience as $(x, y, z, \theta_i, \theta_j)$, where x, y, z are the position coordinates of a reference point on the platform and (θ_i, θ_j) are the joint angles of the Hooke joint attached to the platform. Other coordinates may be chosen.

Assume that the centers of the joints located on the base and on the platform are located on circles with radii R_b and R_p, respectively. A fixed reference frame $O - xyz$ is attached to the base of the mechanism and a moving coordinate frame $P - x'y'z'$ is attached to the platform. In Fig. 5.7, the points of attachment of the actuated legs to the base are represented with B_i and the points of attachment of all legs to the platform are represented by P_i, with $i = 1, \ldots, n$. Point P is the reference point on the platform and its position coordinates are $P(x, y, z)$.

The Cartesian coordinates of the platform are given by the position of point P with respect to the fixed frame, and the orientation of the platform (orientation of frame $P - x'y'z'$ with respect to the fixed frame), represented by matrix $\mathbf{Q}$.

If the coordinates of the point P_i in the moving reference frame are represented with (x_i', y_i', z_i') and the coordinates of the point B_i in the fixed frame are represented by vector $\mathbf{b}_i$, then for $i = 1, \ldots, n$, one has

$$\mathbf{p}_i = \begin{bmatrix} x_i \\ y_i \\ z_i \end{bmatrix}, \quad \mathbf{r}_i' = \begin{bmatrix} x_i' \\ y_i' \\ z_i' \end{bmatrix}, \quad \mathbf{p} = \begin{bmatrix} x \\ y \\ z \end{bmatrix}, \quad \mathbf{b}_i = \begin{bmatrix} b_{ix} \\ b_{iy} \\ b_{iz} \end{bmatrix}, \tag{5.12}$$

where $\mathbf{p}_i$ is the position vector of point P_i expressed in the fixed coordinate frame whose coordinates are defined as (x_i, y_i, z_i), $\mathbf{r}_i'$ is the position vector of point P_i expressed in the moving coordinate frame, and $\mathbf{p}$ is the position vector of point P expressed in the fixed frame as defined above.

One can then write

$$\mathbf{p}_i = \mathbf{p} + \mathbf{Q}\mathbf{r}_\mathrm{i}', \tag{5.13}$$

where the rotation matrix can be written as a function of the n joint angles of the $(n+1)$th leg. This matrix is written as

$$\mathbf{Q} = \mathbf{Q}_0\mathbf{Q}_1 \cdots \mathbf{Q}_n, \quad n = 3, 4, \text{or } 5, \tag{5.14}$$

where $\mathbf{Q}_0$ is the rotation matrix from the fixed reference frame to the first frame (fixed) of the passive constraining leg.

In order to solve the inverse kinematic problem, one must first consider the passive constraining leg as a serial n-dof mechanism whose n Cartesian coordinates are known, which is a well-known problem [10, 72, 84, 94, 95, 102]. Once the solution to the inverse kinematics of this n-dof serial mechanism is found, the complete pose (position and orientation) of the platform can be determined using the direct kinematic equations for this serial mechanism.

Subtracting vector $\mathbf{b}_i$ from both sides of (5.13), one obtains

$$\mathbf{p}_i - \mathbf{b}_i = \mathbf{p} + \mathbf{Q}\mathbf{r}'_i - \mathbf{b}_i, \quad i = 1, \ldots, n, \; n = 3, 4, \text{ or } 5. \tag{5.15}$$

Then, taking the Euclidean norm on both sides of (5.15), one has

$$\| \mathbf{p}_i - \mathbf{b}_i \| = \| \mathbf{p} + \mathbf{Q}\mathbf{r}'_i - \mathbf{b}_i \| = \rho_i, \quad i = 1, \ldots, n, \; n = 3, 4, \text{ or } 5, \tag{5.16}$$

where ρ_i is the length of the ith leg, i.e., the value of the ith joint coordinate. The solution of the inverse kinematic problem for the n-dof platform is therefore completed and can be written as

$$\rho_i^2 = (\mathbf{p}_i - \mathbf{b}_i)^{\mathrm{T}}(\mathbf{p}_i - \mathbf{b}_i), \quad i = 1, \ldots, n, \; n = 3, 4, \text{ or } 5. \tag{5.17}$$

5.3.3 Jacobian Matrices

5.3.3.1 Rigid Mechanisms

The parallel mechanisms studied here comprise two main components, namely, the constraining leg – which can be thought of as a serial mechanism – and the actuated legs acting in parallel.

Considering the constraining leg, one can write

$$\mathbf{J}_{n+1}\dot{\theta}_{n+1} = \mathbf{t}, \quad n = 3, 4, \text{ or } 5, \tag{5.18}$$

where $\mathbf{t} = \left[\omega^{\mathrm{T}} \; \dot{\mathbf{p}}^{\mathrm{T}}\right]^{\mathrm{T}}$ is the twist of the platform, with ω the angular velocity of the platform and

$$\dot{\theta}_{n+1} = \left[\dot{\theta}_{n+1,1} \cdots \dot{\theta}_{n+1,n}\right]^{\mathrm{T}}, \quad n = 3, 4, \text{ or } 5. \tag{5.19}$$

is the joint velocity vector associated with the constraining leg. Matrix $\mathbf{J}_{n+1}$ is the Jacobian matrix of the constraining leg considered as a serial n-dof mechanism.

5.3.3.2 Compliant Model

If the compliance of the links and joints is included, $(6-n)$ virtual joints are added in order to account for the compliance of the links [62]. Hence, the Jacobian matrix of the constraining leg becomes

$$\mathbf{J}'_{n+1}\dot{\theta}'_{n+1} = \mathbf{t}, \quad n = 3, 4, \text{ or } 5, \tag{5.20}$$

where

$$\dot{\theta}'_{n+1} = \left[\dot{\theta}_{n+1,1} \cdots \dot{\theta}_{n+1,6}\right]^{\mathrm{T}}, \quad n = 3, 4, \text{ or } 5. \tag{5.21}$$

5.3.3.3 Global Velocity Equation

Now considering the parallel component of the mechanism, the parallel Jacobian matrix can be obtained by differentiating (5.17) with respect to time, one obtains

$$\rho_i \dot{\rho}_i = (\mathbf{p}_i - \mathbf{b}_i)^{\mathrm{T}} \dot{\mathbf{p}}_i, \quad i = 1, \ldots, n. \tag{5.22}$$

Since one has

$$\dot{\mathbf{Q}} = \Omega \mathbf{Q} \tag{5.23}$$

with

$$\Omega = \mathbf{1} \times \omega = \begin{bmatrix} 0 & -\omega_3 & \omega_2 \\ \omega_3 & 0 & -\omega_1 \\ -\omega_2 & \omega_1 & 0 \end{bmatrix} \tag{5.24}$$

differentiating (5.13), one obtains

$$\dot{\mathbf{p}}_i = \dot{\mathbf{p}} + \dot{\mathbf{Q}}\mathbf{r}'_i. \tag{5.25}$$

Then, for $n = 3, 4$, or 5, (5.22) can be rewritten as

$$\begin{aligned} \rho_i \dot{\rho}_i &= (\mathbf{p}_i - \mathbf{b}_i)^{\mathrm{T}} (\dot{\mathbf{p}} + \dot{\mathbf{Q}}\mathbf{r}'_i) \\ &= (\mathbf{p}_i - \mathbf{b}_i)^{\mathrm{T}} (\dot{\mathbf{p}} + \Omega \mathbf{Q}\mathbf{r}'_i) \\ &= (\mathbf{p}_i - \mathbf{b}_i)^{\mathrm{T}} \dot{\mathbf{p}} + (\mathbf{p}_i - \mathbf{b}_i)^{\mathrm{T}} \Omega \mathbf{Q}\mathbf{r}'_i \\ &= (\mathbf{p}_i - \mathbf{b}_i)^{\mathrm{T}} \dot{\mathbf{p}} + (\mathbf{p}_i - \mathbf{b}_i)^{\mathrm{T}} [\omega \times (\mathbf{Q}\mathbf{r}'_i)] \\ &= (\mathbf{p}_i - \mathbf{b}_i)^{\mathrm{T}} \dot{\mathbf{p}} + [(\mathbf{Q}\mathbf{r}'_i) \times (\mathbf{p}_i - \mathbf{b}_i)]^{\mathrm{T}} \omega, \quad i = 1, \ldots, n. \end{aligned} \tag{5.26}$$

Hence, one can write the velocity equation as

$$\mathbf{A}\mathbf{t} = \mathbf{B}\dot{\rho}, \tag{5.27}$$

where vector $\dot{\rho}$ is defined as

$$\dot{\rho} = \left[\dot{\rho}_1 \; \dot{\rho}_2 \cdots \dot{\rho}_n\right]^{\mathrm{T}}. \tag{5.28}$$

and

$$\mathbf{A} = \begin{bmatrix} \mathbf{m}_1^{\mathrm{T}} \\ \mathbf{m}_2^{\mathrm{T}} \\ \vdots \\ \mathbf{m}_n^{\mathrm{T}} \end{bmatrix}, \quad \mathbf{B} = diag[\rho_1, \rho_2, \ldots, \rho_n], \tag{5.29}$$

where $\mathbf{m}_i$ is a vector with six components, which can be expressed as

$$\mathbf{m}_i = \begin{bmatrix} (\mathbf{Q}\mathbf{r}'_i) \times (\mathbf{p}_i - \mathbf{b}_i) \\ (\mathbf{p}_i - \mathbf{b}_i) \end{bmatrix}. \tag{5.30}$$

Hence, (5.18) or (5.20) relates the twist of the platform to the joint velocities of the passive constraining leg through the serial Jacobian matrix $\mathbf{J}_{n+1}$ or $\mathbf{J}'_{n+1}$ while (5.27) relates the twist of the platform to the actuator velocities through parallel Jacobian matrices $\mathbf{A}$ and $\mathbf{B}$. It should be pointed out that the dimensions of matrix $\mathbf{J}_{n+1}$ will be $(6 \times n)$, matrix $\mathbf{J}'_{n+1}$ will be (6×6), matrix $\mathbf{A}$ will be $(n \times 6)$ and matrix $\mathbf{B}$ will be $(n \times n)$. The derivation of the relationship between Cartesian velocities and joint rates is thereby completed.

5.3.4 Kinetostatic Model for the Mechanism with Rigid Links

In this section, the velocity equations derived in Sect. 5.3.3.3 will be used to obtain the kinetostatic model for the mechanism with rigid links.

According to the principle of virtual work, one has

$$\tau^{\mathrm{T}}\dot{\rho} = \mathbf{w}^{\mathrm{T}}\mathbf{t}, \tag{5.31}$$

where τ is the vector of actuator forces applied at each actuated joint and $\mathbf{w}$ is the wrench (torque and force) applied to the platform and where it is assumed that no gravitational forces act on any of the intermediate links. In practice, gravitational forces may often be neglected in machine tool applications.

One has $\mathbf{w} = \begin{bmatrix} \mathbf{n}^{\mathrm{T}} & \mathbf{f}^{\mathrm{T}} \end{bmatrix}^{\mathrm{T}}$ where $\mathbf{n}$ and $\mathbf{f}$ are, respectively, the external torque and force applied to the platform.

Rearranging (5.27) and substituting it into (5.31), one obtains

$$\tau^{\mathrm{T}}\mathbf{B}^{-1}\mathbf{A}\mathbf{t} = \mathbf{w}^{\mathrm{T}}\mathbf{t}. \tag{5.32}$$

Now, substituting (5.18) into eq. (5.32), one has

$$\tau^{\mathrm{T}}\mathbf{B}^{-1}\mathbf{A}\mathbf{J}_{n+1}\dot{\theta}_{n+1} = \mathbf{w}^{\mathrm{T}}\mathbf{J}_{n+1}\dot{\theta}_{n+1}. \tag{5.33}$$

The latter equation must be satisfied for arbitrary values of $\dot{\theta}_{n+1}$ and hence one can write

$$(\mathbf{A}\mathbf{J}_{n+1})^{\mathrm{T}}\mathbf{B}^{-\mathrm{T}}\tau = \mathbf{J}_{n+1}{}^{\mathrm{T}}\mathbf{w}. \tag{5.34}$$

The latter equation relates the actuator forces to the Cartesian wrench, $\mathbf{w}$, applied at the end-effector in static mode. Since all links are assumed rigid, the compliance of the mechanism will be induced solely by the compliance of the actuators. An actuator compliance matrix $\mathbf{C}$ is therefore defined as

$$\mathbf{C}\tau = \Delta\rho, \tag{5.35}$$

where τ is the vector of actuated joint forces and $\Delta\rho$ is the induced joint displacement. Matrix $\mathbf{C}$ is a $(n \times n)$ diagonal matrix whose ith diagonal entry is the compliance of the ith actuator.

Now, (5.34) can be rewritten as

$$\tau = \mathbf{B}^{\mathrm{T}}(\mathbf{A}\mathbf{J}_{n+1})^{-\mathrm{T}}\mathbf{J}_{n+1}{}^{\mathrm{T}}\mathbf{w}. \tag{5.36}$$

The substitution of (5.36) into (5.35) then leads to

$$\Delta\rho = \mathbf{C}\mathbf{B}^{\mathrm{T}}(\mathbf{A}\mathbf{J}_{n+1})^{-\mathrm{T}}\mathbf{J}_{n+1}{}^{\mathrm{T}}\mathbf{w}. \tag{5.37}$$

Moreover, for a small displacement vector $\Delta\rho$, (5.27) can be written as

$$\Delta\rho \simeq \mathbf{B}^{-1}\mathbf{A}\Delta\mathbf{c}, \tag{5.38}$$

where $\Delta\mathbf{c}$ is a vector of small Cartesian displacement and rotation defined as

$$\Delta\mathbf{c} = \left[\, \Delta\mathbf{p}^{\mathrm{T}} \;\; \Delta\alpha^{\mathrm{T}} \,\right]^{\mathrm{T}} \tag{5.39}$$

in which $\Delta\alpha$, the change of orientation, is defined from (5.23) and (5.24) as

$$\Delta\alpha = \mathrm{vect}(\Delta\mathbf{Q}\mathbf{Q}^{\mathrm{T}}), \tag{5.40}$$

where $\Delta\mathbf{Q}$ is the variation of the rotation matrix and $\mathrm{vect}(\cdot)$ is the vector linear invariant of its matrix argument.

Similarly, (5.18) can also be written, for small displacements, as

$$\mathbf{J}_{n+1}\Delta\theta_{n+1} \simeq \Delta\mathbf{c}, \tag{5.41}$$

where $\Delta\theta_{n+1}$ is a vector of small variations of the joint coordinates of the constraining leg.

Substituting (5.38) into (5.37), one obtains

$$\mathbf{B}^{-1}\mathbf{A}\Delta\mathbf{c} = \mathbf{C}\mathbf{B}^{\mathrm{T}}(\mathbf{A}\mathbf{J}_{n+1})^{-\mathrm{T}}\mathbf{J}_{n+1}^{\mathrm{T}}\mathbf{w}. \tag{5.42}$$

Premultiplying both sides of (5.42) by $\mathbf{B}$, and substituting (5.41) into (5.42), one obtains,

$$\mathbf{A}\mathbf{J}_{n+1}\Delta\theta_{n+1} = \mathbf{B}\mathbf{C}\mathbf{B}^{\mathrm{T}}(\mathbf{A}\mathbf{J}_{n+1})^{-\mathrm{T}}\mathbf{J}_{n+1}^{\mathrm{T}}\mathbf{w}. \tag{5.43}$$

Then, premultiplying both sides of (5.43) by $(\mathbf{A}\mathbf{J}_{n+1})^{-1}$, one obtains,

$$\Delta\theta_{n+1} = (\mathbf{A}\mathbf{J}_{n+1})^{-1}\mathbf{B}\mathbf{C}\mathbf{B}^{\mathrm{T}}(\mathbf{A}\mathbf{J}_{n+1})^{-\mathrm{T}}\mathbf{J}_{n+1}^{\mathrm{T}}\mathbf{w} \tag{5.44}$$

and finally, premultiplying both sides of (5.44) by $\mathbf{J}_{n+1}$, one obtains,

$$\Delta\mathbf{c} = \mathbf{J}_{n+1}(\mathbf{A}\mathbf{J}_{n+1})^{-1}\mathbf{B}\mathbf{C}\mathbf{B}^{\mathrm{T}}(\mathbf{A}\mathbf{J}_{n+1})^{-\mathrm{T}}\mathbf{J}_{n+1}^{\mathrm{T}}\mathbf{w}. \tag{5.45}$$

Hence, one obtains the Cartesian compliance matrix as

$$\mathbf{C}_c = \mathbf{J}_{n+1}(\mathbf{A}\mathbf{J}_{n+1})^{-1}\mathbf{B}\mathbf{C}\mathbf{B}^T(\mathbf{A}\mathbf{J}_{n+1})^{-T}\mathbf{J}_{n+1}^T \tag{5.46}$$

with

$$\Delta\mathbf{c} = \mathbf{C}_c\mathbf{w}, \tag{5.47}$$

where $\mathbf{C}_c$ is a symmetric positive semidefinite (6×6) matrix, as expected.

It is pointed out that, in nonsingular configurations, the rank of $\mathbf{B}$, $\mathbf{C}$, and $\mathbf{J}_{n+1}$ is n, and hence the rank of $\mathbf{C}_c$ will be n, where $n = 3, 4$, or 5, depending on the degree of freedom of the mechanism. Hence, the nullspace of matrix $\mathbf{C}_c$ will not be empty and there will exist a set of vectors $\mathbf{w}$ that will induce no Cartesian displacement $\Delta\mathbf{c}$. This corresponds to the wrenches that are supported by the constraining leg, which is considered infinitely rigid. These wrenches are orthogonal complements of the allowable twists at the platform. Hence, matrix $\mathbf{C}_c$ cannot be inverted and this is why it was more convenient to use compliance matrices rather than stiffness matrices in the above derivation.

In Sect. 5.3.5, the kinetostatic model will be rederived for the case in which the flexibility of the links is considered. In this case, stiffness matrices will be used.

5.3.5 Kinetostatic Model for the Mechanism with Flexible Links

According to the principle of virtual work, one can write

$$\mathbf{w}^T\mathbf{t} = \tau_{n+1}^T\dot{\theta}'_{n+1} + \tau^T\dot{\rho}, \tag{5.48}$$

where τ is the vector of actuator forces and $\dot{\rho}$ is the vector of actuator velocities (actuated legs), and τ_{n+1} is the vector of joint torques in the constraining leg. This vector is defined as follows, where $\mathbf{K}_{n+1}$ is the stiffness matrix of the constraining leg,

$$\tau_{n+1} = \mathbf{K}_{n+1}\Delta\theta'_{n+1}. \tag{5.49}$$

Matrix $\mathbf{K}_{n+1}$ is a diagonal (6×6) matrix in which the ith diagonal entry is zero if it is associated with a real joint while it is equal to k_i if it is associated with a virtual joint, where k_i is the stiffness of the virtual spring located at the ith joint. The stiffness of the virtual springs is determined using the structural properties of the flexible links as shown in Chap. 5.

From (5.20) and (5.27), (5.48) can be rewritten as

$$\mathbf{w}^T\mathbf{t} = \tau_{n+1}^T(\mathbf{J}'_{n+1})^{-1}\mathbf{t} + \tau^T\mathbf{B}^{-1}\mathbf{A}\mathbf{t}. \tag{5.50}$$

Since this equation is valid for any value of $\mathbf{t}$, one can write

$$\mathbf{w} = (\mathbf{J}'_{n+1})^{-T}\tau_{n+1} + \mathbf{A}^T\mathbf{B}^{-T}\tau, \tag{5.51}$$

which can be rewritten as

$$\mathbf{w} = (\mathbf{J}'_{n+1})^{-\mathrm{T}}\mathbf{K}_{n+1}\Delta\theta'_{n+1} + \mathbf{A}^{\mathrm{T}}\mathbf{B}^{-\mathrm{T}}\mathbf{K}_{\mathrm{J}}\Delta\rho, \tag{5.52}$$

where $\mathbf{K}_J$ is a $(n \times n)$ diagonal joint stiffness matrix for the actuated joints.

Using the kinematic equations, one can then write:

$$\mathbf{w} = (\mathbf{J}'_{n+1})^{-\mathrm{T}}\mathbf{K}_{n+1}(\mathbf{J}'_{n+1})^{-1}\Delta\mathbf{c} + \mathbf{A}^{\mathrm{T}}\mathbf{B}^{-\mathrm{T}}\mathbf{K}_{\mathrm{J}}\mathbf{B}^{-1}\mathbf{A}\Delta\mathbf{c} \tag{5.53}$$

which is in the form

$$\mathbf{w} = \mathbf{K}\Delta\mathbf{c}, \tag{5.54}$$

where $\mathbf{K}$ is the Cartesian stiffness matrix, which is equal to

$$\mathbf{K} = [(\mathbf{J}'_{n+1})^{-\mathrm{T}}\mathbf{K}_{n+1}(\mathbf{J}'_{n+1})^{-1} + \mathbf{A}^{\mathrm{T}}\mathbf{B}^{-\mathrm{T}}\mathbf{K}_{\mathrm{J}}\mathbf{B}^{-1}\mathbf{A}] \tag{5.55}$$

Matrix $\mathbf{K}$ is a symmetric (6×6) positive semidefinite matrix, as expected. However, in this case, matrix $\mathbf{K}$ will be of full rank in nonsingular configurations. Indeed, the sum of the two terms in (5.55) will span the complete space of constraint wrenches.

5.3.6 *Examples*

5.3.6.1 5-dof Parallel Mechanism

This mechanism is illustrated in Fig. 5.6, the compliance matrix for the mechanism with rigid links can be written as

$$\mathbf{C}_{\mathrm{c}} = \mathbf{J}_6(\mathbf{A}\mathbf{J}_6)^{-1}\mathbf{B}\mathbf{C}\mathbf{B}^{\mathrm{T}}(\mathbf{A}\mathbf{J}_6)^{-\mathrm{T}}\mathbf{J}_6^{\mathrm{T}}, \tag{5.56}$$

where

$$\mathbf{C} = \mathrm{diag}[c_1, c_2, c_3, c_4, c_5], \tag{5.57}$$

where c_i is the stiffness of the ith actuator and $\mathbf{J}_6$ is the Jacobian matrix of the rigid constraining leg in this 5-dof case. Matrices $\mathbf{A}$ and $\mathbf{B}$ are the Jacobian matrices of the structure without the passive constraining leg.

Similarly, the stiffness matrix for the mechanism with flexible links can be written as

$$\mathbf{K} = [(\mathbf{J}'_6)^{-\mathrm{T}}\mathbf{K}_6(\mathbf{J}'_6)^{-1} + \mathbf{A}^{\mathrm{T}}\mathbf{B}^{-\mathrm{T}}\mathbf{K}_{\mathrm{J}}\mathbf{B}^{-1}\mathbf{A}], \tag{5.58}$$

where

$$\mathbf{K}_6 = \mathrm{diag}[0, k_{62}, 0, 0, 0, 0], \tag{5.59}$$

Table 5.4 Cartesian compliance of the 5-dof mechanism with flexible links and with rigid links

K_a	K_{passive}	κ_{θ_x}	κ_{θ_y}	κ_{θ_z}	κ_x	κ_y	κ_z
1,000	1,000	0.0862844	0.0980888	0.261173	0.0735117	0.0303542	0.000255386
1,000	$10^1 K_a$	0.0862729	0.098054	0.259051	0.0734333	0.0303281	0.000255386
1,000	$10^2 K_a$	0.0862717	0.0980506	0.258838	0.0734254	0.0303255	0.000255386
1,000	$10^3 K_a$	0.0862716	0.0980502	0.258817	0.0734246	0.0303252	0.000255386
1,000	$10^4 K_a$	0.0862716	0.0980502	0.258815	0.0734245	0.0303252	0.000255386
1,000	$10^5 K_a$	0.0862716	0.0980502	0.258815	0.0734245	0.0303252	0.000255386
1,000	Rigid	0.0862716	0.0980502	0.258815	0.0734245	0.0303252	0.000255386

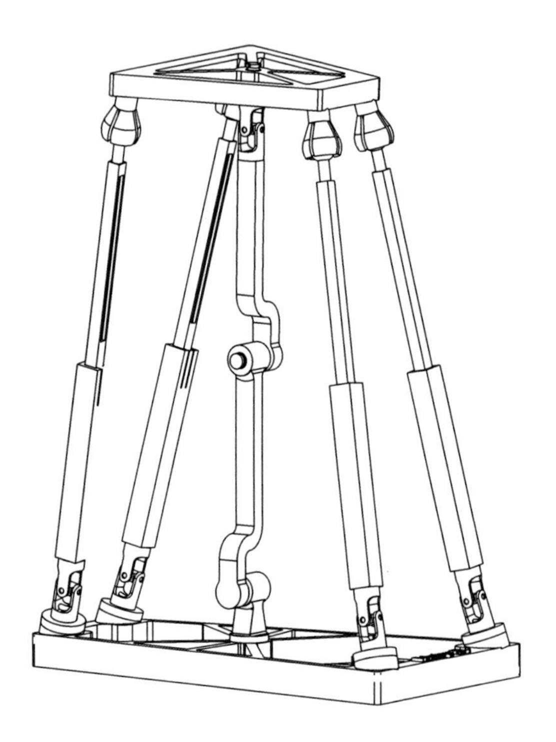

Fig. 5.8 CAD model of the spatial 4-dof parallel manipulator with prismatic actuators (Figure by Gabriel Coté)

where k_{62} is the stiffness of the virtual joint and $\mathbf{J}_6'$ is the Jacobian matrix of the compliant passive constraining leg in this 5-dof case, while **A** and **B** are the Jacobian matrices of the structure without the special leg.

The comparison between the mechanism with rigid links (without virtual joints) and the mechanism with flexible links (with virtual joints) is given in Table 5.4.

5.3.6.2 4-dof Parallel Mechanism

This mechanism is illustrated in Fig. 5.8, the compliance matrix for the mechanism with rigid links can be written as

$$\mathbf{C}_c = \mathbf{J}_5(\mathbf{A}\mathbf{J}_5)^{-1}\mathbf{B}\mathbf{C}\mathbf{B}^{\mathrm{T}}(\mathbf{A}\mathbf{J}_5)^{-\mathrm{T}}\mathbf{J}_5^{\mathrm{T}}, \tag{5.60}$$

Table 5.5 Cartesian compliance of the 4-dof mechanism with flexible links and with rigid links

K_{actuator}	K_{passive}	κ_{θ_x}	κ_{θ_y}	κ_{θ_z}	κ_x	κ_y	κ_z
1,000	1,000	0.52371	1.41939	1.5×10^{-3}	0.915208	5.78×10^{-4}	0.0111974
1,000	$10^1 K_a$	0.523128	1.41007	1.5×10^{-4}	0.912371	5.78×10^{-5}	0.0111751
1,000	$10^2 K_a$	0.51707	1.40514	1.5×10^{-5}	0.909087	5.78×10^{-6}	0.0111429
1,000	$10^3 K_a$	0.516464	1.40464	1.5×10^{-6}	0.908758	5.78×10^{-7}	0.0111396
1,000	$10^4 K_a$	0.516404	1.4046	1.5×10^{-7}	0.908726	5.78×10^{-8}	0.0111393
1,000	$10^5 K_a$	0.516398	1.40459	1.5×10^{-8}	0.908722	5.78×10^{-9}	0.0111393
1,000	$10^6 K_a$	0.516397	1.40459	1.5×10^{-9}	0.908722	5.78×10^{-10}	0.0111393
1,000	$10^7 K_a$	0.516397	1.40459	1.5×10^{-10}	0.908722	5.78×10^{-11}	0.0111393
1,000	Rigid	0.516397	1.40459	0.0	0.908722	0.0	0.0111393

where

$$\mathbf{C} = \text{diag}[c_1, c_2, c_3, c_4] \tag{5.61}$$

with c_i is the compliance of the ith actuator and $\mathbf{J}_5$ is the Jacobian matrix of the rigid constraining leg in this 4-dof case. Matrices $\mathbf{A}$ and $\mathbf{B}$ are the Jacobian matrices of the structure without the constraining leg.

Similarly, the stiffness matrix for the mechanism with flexible links can be written as

$$\mathbf{K} = [(\mathbf{J}_5')^{-\mathrm{T}}\mathbf{K}_5(\mathbf{J}_5')^{-1} + \mathbf{A}^{\mathrm{T}}\mathbf{B}^{-\mathrm{T}}\mathbf{K}_{\mathrm{J}}\mathbf{B}^{-1}\mathbf{A}], \tag{5.62}$$

where

$$\mathbf{K}_5 = \text{diag}[0, k_{52}, 0, k_{54}, 0, 0] \tag{5.63}$$

with k_{52} and k_{54} are the stiffnesses of the virtual joints. Matrix $\mathbf{J}_5'$ is the Jacobian matrix of the compliant constraining leg in this 4-dof case, while $\mathbf{A}$ and $\mathbf{B}$ are the Jacobian matrices of the structure without the constraining leg.

The compliance comparison between the mechanism with rigid links (without virtual joints) and the mechanism with flexible links (with virtual joints) is given in Table 5.5. Again, the effect of the link flexibility is clearly demonstrated.

5.3.6.3 3-dof Parallel Mechanism

This mechanism is illustrated in Fig. 5.9, the compliance matrix for the rigid mechanism can be written as

$$\mathbf{C}_{\mathrm{c}} = \mathbf{J}_4(\mathbf{A}\mathbf{J}_4)^{-1}\mathbf{B}\mathbf{C}\mathbf{B}^{\mathrm{T}}(\mathbf{A}\mathbf{J}_4)^{-\mathrm{T}}\mathbf{J}_4^{\mathrm{T}}, \tag{5.64}$$

where $\mathbf{C} = \text{diag}[c_1, c_2, c_3]$, with c_1, c_2, and c_3 the compliance of the actuators and $\mathbf{J}_4$ is the Jacobian matrix of the constraining leg in this 3-dof case. Matrices $\mathbf{A}$ and $\mathbf{B}$ are the Jacobian matrices of the structure without the special leg.

Similarly, the stiffness matrix for the mechanism with flexible links can be written as

$$\mathbf{K} = [(\mathbf{J}_4')^{-\mathrm{T}}\mathbf{K}_4(\mathbf{J}_4')^{-1} + \mathbf{A}^{\mathrm{T}}\mathbf{B}^{-\mathrm{T}}\mathbf{K}_{\mathrm{J}}\mathbf{B}^{-1}\mathbf{A}], \tag{5.65}$$

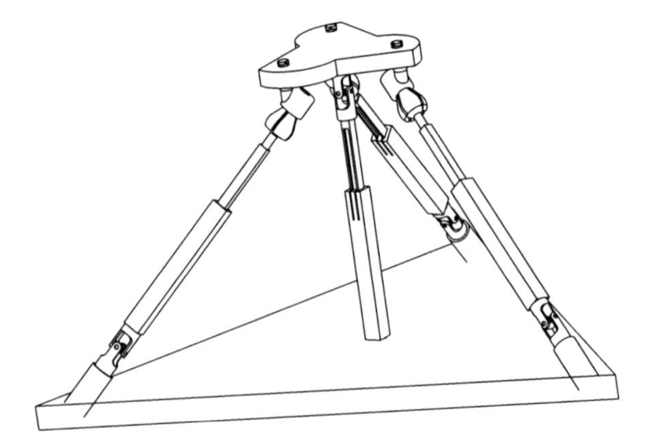

Fig. 5.9 CAD model of the spatial 3-dof parallel mechanism with prismatic actuators (Figure by Gabriel Coté)

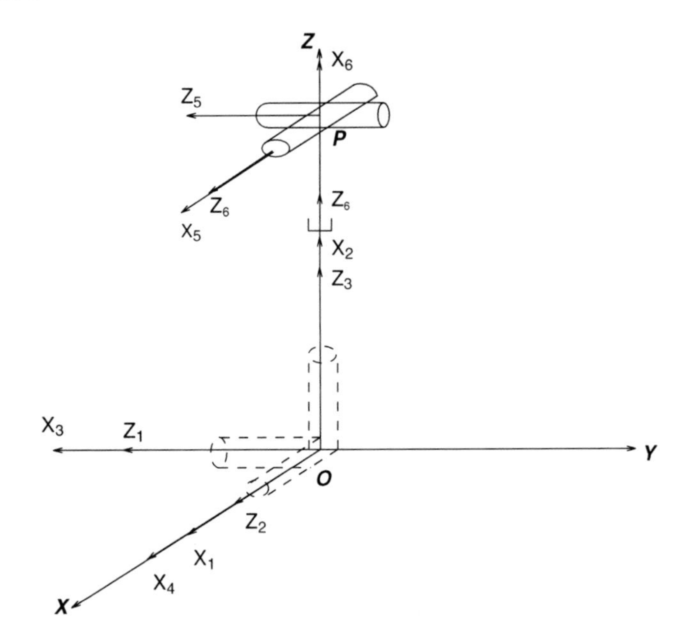

Fig. 5.10 The passive constraining leg with flexible links

where

$$\mathbf{K}_4 = diag[k_{41}, k_{42}, k_{43}, 0, 0, 0], \tag{5.66}$$

where k_{41}, k_{42}, and k_{43} are the stiffnesses of the virtual joints introduced to account for the flexibility of the links in the constraining leg. The architecture of the constraining leg including the virtual joints is represented in Fig. 5.10, and $\mathbf{J}_4'$ is the Jacobian matrix of the constraining leg in this 3-dof case, while $\mathbf{A}$ and $\mathbf{B}$ are the Jacobian matrices of the structure without the constraining leg.

Table 5.6 Comparison of the mechanism compliance between the mechanism with rigid links and the mechanism with flexible links

K_a	K_{passive}	$\kappa_{\theta_x} = \kappa_{\theta_y}$	κ_{θ_z}	$\kappa_x = \kappa_y$	κ_z
1,000	1,000	0.192784	10^{-3}	4.624×10^{-4}	3.4569×10^{-4}
1,000	$10K_a$	0.192509	10^{-4}	4.624×10^{-5}	3.4567×10^{-4}
1,000	10^2K_a	0.192081	10^{-5}	4.624×10^{-6}	3.4566×10^{-4}
1,000	10^3K_a	0.192038	10^{-6}	4.624×10^{-7}	3.4566×10^{-4}
1,000	10^4K_a	0.192034	10^{-7}	4.624×10^{-8}	3.4566×10^{-4}
1,000	10^5K_a	0.192034	10^{-8}	4.624×10^{-9}	3.4566×10^{-4}
1,000	10^6K_a	0.192034	10^{-9}	4.624×10^{-10}	3.4566×10^{-4}
1,000	10^7K_a	0.192034	10^{-10}	4.624×10^{-11}	3.4566×10^{-4}
1,000	Rigid	0.192034	0.0	0.0	3.4566×10^{-4}

The comparison between the mechanism with rigid links (without virtual joints) and the mechanism with flexible links (with virtual joints) is given in Table 5.6. The Cartesian compliance in each of the directions is given for a reference configuration of the mechanism, for progressively increasing values of the link stiffnesses.

From Table 5.6, one can find that with the improvement of the link stiffness, the mechanism's compliance is very close to that of mechanism with rigid link, this means that one can consider the flexible mechanism as a rigid one only if the link stiffness reaches a high value. This indicates that one cannot neglect the effects of link flexibility. It is also applicable to practical application. An industrial example of Tricept machine tools is presented and analyzed in Zhang and Gosselin [101].

5.4 Conclusions

A 6-dof fully parallel robot manipulator with prismatic actuators has been introduced and analyzed using kinetostatic method. Then, a new family of n-DOF parallel mechanisms with one passive constraining leg has been introduced in this chapter. This type of architecture can be used in several applications including machine tools. The kinematic analysis of this family of spatial parallel ndof mechanisms has been presented. The geometric configurations of the mechanisms are shown. In this chapter, only mechanisms with prismatic actuators have been discussed. Solutions for the inverse kinematic problem have been given. The Jacobian matrices obtained have been used to establish the kinetostatic model of the mechanisms. The lumped link and joint compliances have been used for the study of the Cartesian compliance. Finally, examples have been investigated and numerical results have been obtained. The results clearly demonstrate the relevance of the kinetostatic analysis in the context of design of such mechanisms.

Additionally, there are some common design guidelines for this series of mechanisms, they are:

1. With the improvement of link stiffness, the mechanism's compliance is very close to that of mechanism with rigid link, this suggests that we can assume the flexible mechanism to be rigid only if the link stiffness reaches a high value ($10^7 K_a$).
2. The passive constraining leg's lumped stiffness does not affect all directional stiffnesses, it only plays the role of limiting the platform's motion to the desired ones.
3. The limitation of the platform's degree of freedom is dependent on the actuator stiffness and link stiffness.
4. If the passive constraining leg's lumped stiffness is fixed as the same value as actuator's (K_a), then it cannot adequately limit the motion to the desired degree-of-freedom, only if the passive constraining leg's lumped stiffness is large enough ($10^2 K_a$), then it begins to efficiently play the role of limiting the platform motion to the desired ones.

Chapter 6
Spatial Parallel Robotic Machines with Revolute Actuators

6.1 Preamble

In this chapter, first, a six degrees of freedom fully parallel robotic machine with revolute actuators is presented and analyzed. Then, a serial of parallel manipulators with 3-dof, 4-dof, and 5-dof whose degree of freedom is dependent on an additional passive leg, this passive leg is connecting the center between the base and the moving platform. Together with the inverse kinematics and velocity equations for both rigid-link and flexible-link mechanisms, a general kinetostatic model is established for the analysis of the structural rigidity and accuracy of this family of mechanisms, case studies for 3-dof, 4-dof, and 5-dof mechanisms are given in detail to illustrate the results.

6.2 Six Degrees of Freedom Parallel Robotic Machine with Revolute Actuators

6.2.1 Geometric Modeling

Figures 6.1 and 6.2 represent a 6-dof parallel mechanism with revolute actuators. This mechanism consists of six actuated legs with identical topology, connecting the fixed base to a moving platform. The kinematic chains consist – from base to platform – of an actuated revolute joint, a moving link, a Hooke joint, a second moving link, and a spherical joint attached to the platform. A fixed reference frame $O - xyz$ is connected to the base of the mechanism and a moving coordinate frame $P - x'y'z'$ is connected to the platform.

D. Zhang, *Parallel Robotic Machine Tools*, DOI 10.1007/978-1-4419-1117-9_6,

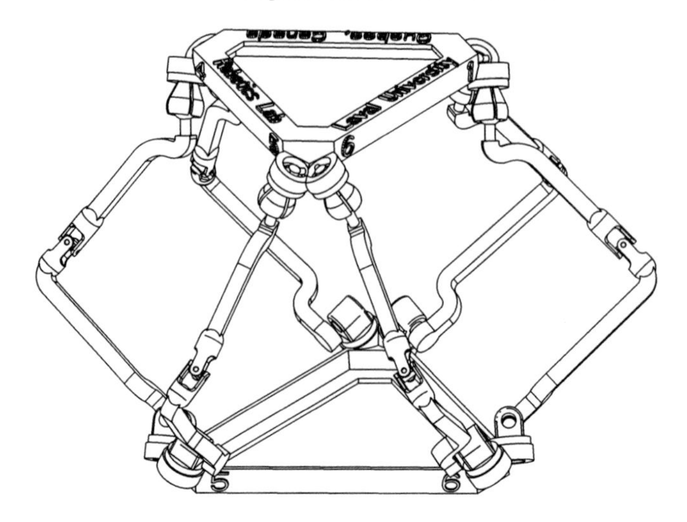

Fig. 6.1 CAD model of the spatial 6-dof parallel mechanism with revolute actuators (Figure by Thierry Laliberté and Gabriel Coté)

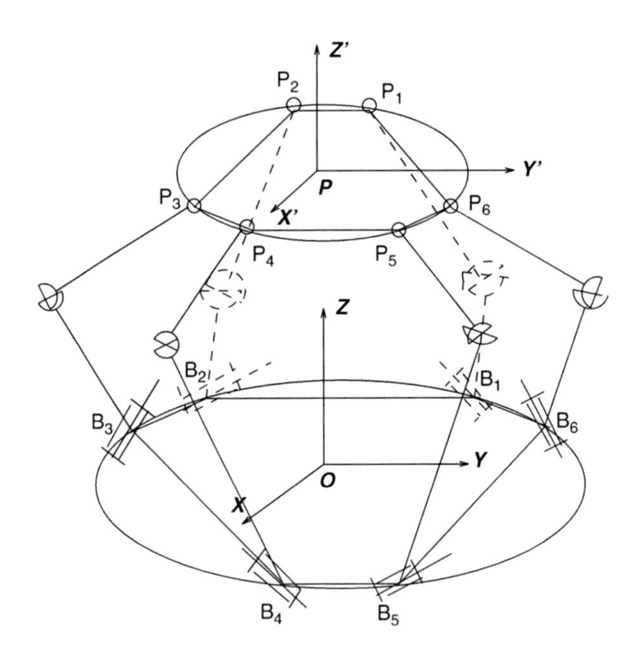

Fig. 6.2 Schematic representation of the spatial 6-dof parallel mechanism with revolute actuators

6.2.2 Global Velocity Equation

6.2.2.1 Rigid Model

The global velocity equation for rigid model can be expressed as

$$\mathbf{At} = \mathbf{B}\dot{\theta}, \tag{6.1}$$

where vectors $\dot{\theta}$ and $\mathbf{t}$ are defined as

$$\dot{\theta} = \left[\dot{\theta}_1 \cdots \dot{\theta}_6\right]^{\mathrm{T}}, \tag{6.2}$$

$$\mathbf{t} = \left[\omega^{\mathrm{T}}, \dot{\mathbf{p}}^{\mathrm{T}}\right]^{\mathrm{T}} \tag{6.3}$$

and

$$\mathbf{A} = \left[\mathbf{m}_1\ \mathbf{m}_2\ \mathbf{m}_3\ \mathbf{m}_4\ \mathbf{m}_5\ \mathbf{m}_6\right]^{\mathrm{T}} \tag{6.4}$$

$$\begin{aligned}\mathbf{B} = {} & \mathrm{diag}[(\mathbf{p}_1 - \mathbf{b}'_1)^{\mathrm{T}}\mathbf{e}_1, (\mathbf{p}_2 - \mathbf{b}'_2)^{\mathrm{T}}\mathbf{e}_2, (\mathbf{p}_3 - \mathbf{b}'_3)^{\mathrm{T}}\mathbf{e}_3, \\ & (\mathbf{p}_4 - \mathbf{b}'_4)^{\mathrm{T}}\mathbf{e}_4, (\mathbf{p}_5 - \mathbf{b}'_5)^{\mathrm{T}}\mathbf{e}_5, (\mathbf{p}_6 - \mathbf{b}'_6)^{T}\mathbf{e}_6]\end{aligned} \tag{6.5}$$

and $\mathbf{m}_i$ is a six-dimensional vector expressed as

$$\mathbf{m}_i = \begin{bmatrix}(\mathbf{Q}\mathbf{r}'_i) \times (\mathbf{p}_i - \mathbf{b}'_i) \\ (\mathbf{p}_i - \mathbf{b}'_i)\end{bmatrix}, \quad i = 1, \ldots, 6 \tag{6.6}$$

and again, the Jacobian matrix $\mathbf{J}$ can be written as

$$\mathbf{J} = \mathbf{B}^{-1}\mathbf{A}. \tag{6.7}$$

6.2.2.2 Compliant Model

For the case of compliant model, one can obtain the global velocity equation as

$$\mathbf{At} = \mathbf{B}\dot{\theta}, \tag{6.8}$$

where vector $\dot{\theta}$ is defined as

$$\dot{\theta} = \left[\dot{\theta}_{11}\ \dot{\theta}_{12}\ \dot{\theta}_{21}\ \dot{\theta}_{22}\ \dot{\theta}_{31}\ \dot{\theta}_{32}\ \dot{\theta}_{41}\ \dot{\theta}_{42}\ \dot{\theta}_{51}\ \dot{\theta}_{52}\ \dot{\theta}_{61}\ \dot{\theta}_{62}\right]^{\mathrm{T}} \tag{6.9}$$

matrix **A** and its terms are as given in (6.4) and (6.6) and

$$\mathbf{B}_{6\times12} = \begin{bmatrix} b_{11} & b_{12} & 0 & 0 & 0 & 0 & 0 & 0 & 0 & 0 & 0 & 0 \\ 0 & 0 & b_{21} & b_{22} & 0 & 0 & 0 & 0 & 0 & 0 & 0 & 0 \\ 0 & 0 & 0 & 0 & b_{31} & b_{32} & 0 & 0 & 0 & 0 & 0 & 0 \\ 0 & 0 & 0 & 0 & 0 & 0 & b_{41} & b_{42} & 0 & 0 & 0 & 0 \\ 0 & 0 & 0 & 0 & 0 & 0 & 0 & 0 & b_{51} & b_{52} & 0 & 0 \\ 0 & 0 & 0 & 0 & 0 & 0 & 0 & 0 & 0 & 0 & b_{61} & b_{62} \end{bmatrix}, \quad (6.10)$$

where

$$b_{ij} = (\mathbf{p}_i - \mathbf{b}_i')^{\mathrm{T}}\mathbf{d}_{ij}, \quad i = 1, \ldots, 6, \; j = 1, 2 \quad (6.11)$$

The derivation of the relationship between Cartesian velocities and joint rates is thereby completed.

6.2.3 Stiffness Model

Again, the stiffness of the structure has been obtained as

$$\mathbf{K} = \mathbf{J}^{\mathrm{T}}\mathbf{K}_{\mathrm{J}}\mathbf{J}. \quad (6.12)$$

One obtains

$$\mathbf{t} = \mathbf{J}'\dot{\theta}, \quad (6.13)$$

where

$$\mathbf{J}' = \mathbf{A}^{-1}\mathbf{B} \quad (6.14)$$

according to the principle of virtual work, one has

$$\tau^{\mathrm{T}}\dot{\theta} = \mathbf{w}^{\mathrm{T}}\mathbf{t}, \quad (6.15)$$

where τ is a vector of the actuator torques applied at each actuated joint or joint with spring. If we assume that no gravitational forces act on any of the intermediate links and **w** is a vector composed of forces and moments (hereafter called wrench) applied by the end-effector. Substituting (6.13) into (6.15) one can obtain

$$\tau = \mathbf{J}'^{\mathrm{T}}\mathbf{w}. \quad (6.16)$$

The joint forces and displacements of each joint can be related by Hooke's law, i.e.,

$$\tau = \mathbf{K}_{\mathrm{J}}\Delta\theta. \quad (6.17)$$

$\Delta\theta$ only includes the actuated joints and joint with springs, i.e.,

$$\mathbf{K}_{\mathrm{J}}\Delta\theta = \mathbf{J}'^{\mathrm{T}}\mathbf{w} \quad (6.18)$$

hence

$$\Delta\theta = \mathbf{K}_{\mathrm{J}}^{-1}\mathbf{J}'^{\mathrm{T}}\mathbf{w}. \tag{6.19}$$

premultiplying by $\mathbf{J}'$ on both sides, one obtains

$$\mathbf{J}'\Delta\theta = \mathbf{J}'\mathbf{K}_{\mathrm{J}}^{-1}\mathbf{J}'^{\mathrm{T}}\mathbf{w}. \tag{6.20}$$

Substituting (6.13) into (6.20), one obtains

$$\mathbf{t} = \mathbf{J}'\mathbf{K}_{\mathrm{J}}^{-1}\mathbf{J}'^{\mathrm{T}}\mathbf{w}, \tag{6.21}$$

therefore, one obtains the compliance matrix of the mechanism κ as follow

$$\kappa = \mathbf{J}'\mathbf{K}_{\mathrm{J}}^{-1}\mathbf{J}'^{\mathrm{T}} \tag{6.22}$$

and the system stiffness matrix is

$$\mathbf{K} = [\mathbf{J}'\mathbf{K}_{\mathrm{J}}^{-1}\mathbf{J}'^{\mathrm{T}}]^{-1}, \tag{6.23}$$

where

$$\mathbf{K}_{\mathrm{J}} = \mathrm{diag}[k_{11}, k_{12}, k_{21}, k_{22}, k_{31}, k_{32}, k_{41}, k_{42}, k_{51}, k_{52}, k_{61}, k_{62}] \tag{6.24}$$

where k_{i1} is stiffness of the ith actuator and k_{i2} is the lumped stiffness of each leg.

In order to illustrate the effect of the flexible links on the parallel mechanism, an example of 6-dof mechanism is presented. The parameters are given as

$$\begin{aligned}
\theta_{\mathrm{p}} &= 22.34°, \theta_{\mathrm{b}} = 42.883°, \\
R_{\mathrm{p}} &= 6\,\mathrm{cm}, R_{\mathrm{b}} = 15\,\mathrm{cm}, \\
l_{i1} &= 46\,\mathrm{cm}, l_{i2} = 36\,\mathrm{cm}, \quad i = 1, \ldots, 6 \\
k_{i1} &= 1{,}000\,\mathrm{Nm}, \quad i = 1, \ldots, 6,
\end{aligned}$$

where k_{i1} is the stiffness of each leg, l_{i1}, l_{i2} are the link lengths for the 1st and 2nd link of each leg, and the Cartesian coordinates are given by

$$\begin{aligned}
x \in [-3, 3]\,\mathrm{cm}, y \in [-3, 3]\,\mathrm{cm}, z = 68\,\mathrm{cm}, \\
\phi = 0, \theta = 0, \psi = 0.
\end{aligned}$$

Figure 6.3 shows the variation of the stiffness for the above example. The comparison between the parallel mechanism with rigid link and the parallel mechanism with flexible links is given in Table 6.1. The results are similar to those obtained in previous cases.

From Fig. 6.4, one can find that K_x and K_y, K_{θ_x} and K_{θ_y} are symmetric with respect to each other. In Fig. 6.4(a), the stiffness in X becomes higher when the

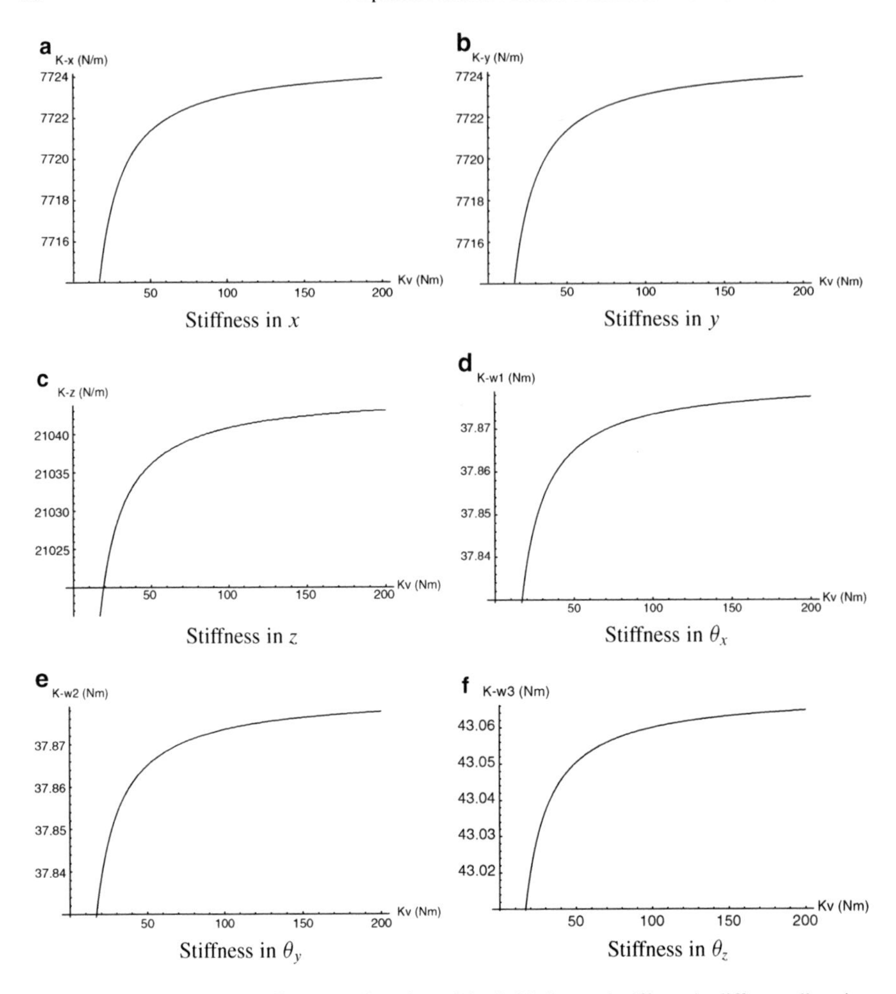

Fig. 6.3 Evolution of the stiffness as a function of the link's lumped stiffness in different directions

Table 6.1 Comparison of the mechanism stiffness between the mechanism with rigid links and the mechanism with flexible links

K_{actuator}	K_{link}	K_x	K_y	K_z	K_{θ_x}	K_{θ_y}	K_{θ_z}
1,000	1,000	3,700.65	3,700.65	10,082.1	18.1478	18.1478	20.633
1,000	$10^1 K_a$	6,967.15	6,967.15	18,981.4	34.1665	34.1665	38.8454
1,000	$10^2 K_a$	7,641.67	7,641.67	20,819.1	37.4743	37.4743	42.6062
1,000	$10^3 K_a$	7,716.37	7,716.37	21,022.6	37.8406	37.8406	43.0227
1,000	$10^4 K_a$	7,723.92	7,723.92	21,043.2	37.8777	37.8777	43.0648
1,000	$10^5 K_a$	7,724.68	7,724.68	21,045.2	37.8814	37.8814	43.069
1,000	$10^6 K_a$	7,724.76	7,724.76	21,045.4	37.8818	37.8818	43.0694
1,000	$10^7 K_a$	7,724.76	7,724.76	21,045.4	37.8818	37.8818	43.0695
1,000	Rigid	7,724.76	7,724.76	21,045.4	37.8818	37.8818	43.0695

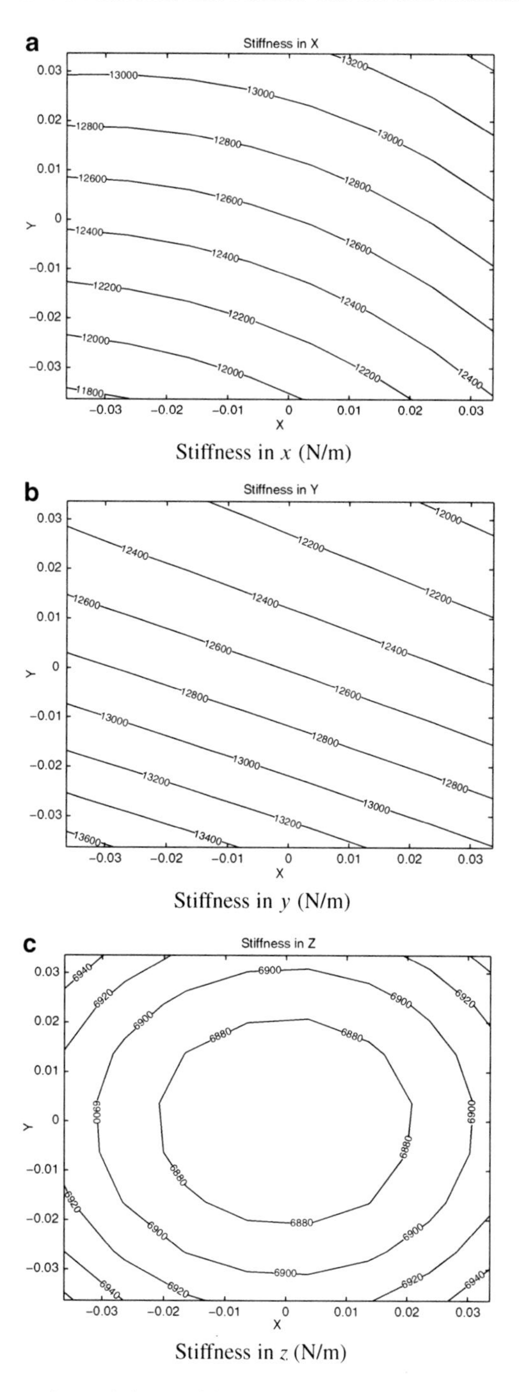

Fig. 6.4 Stiffness mappings of the spatial 6-dof parallel mechanism with revolute actuators (6 legs) (all length units in m)

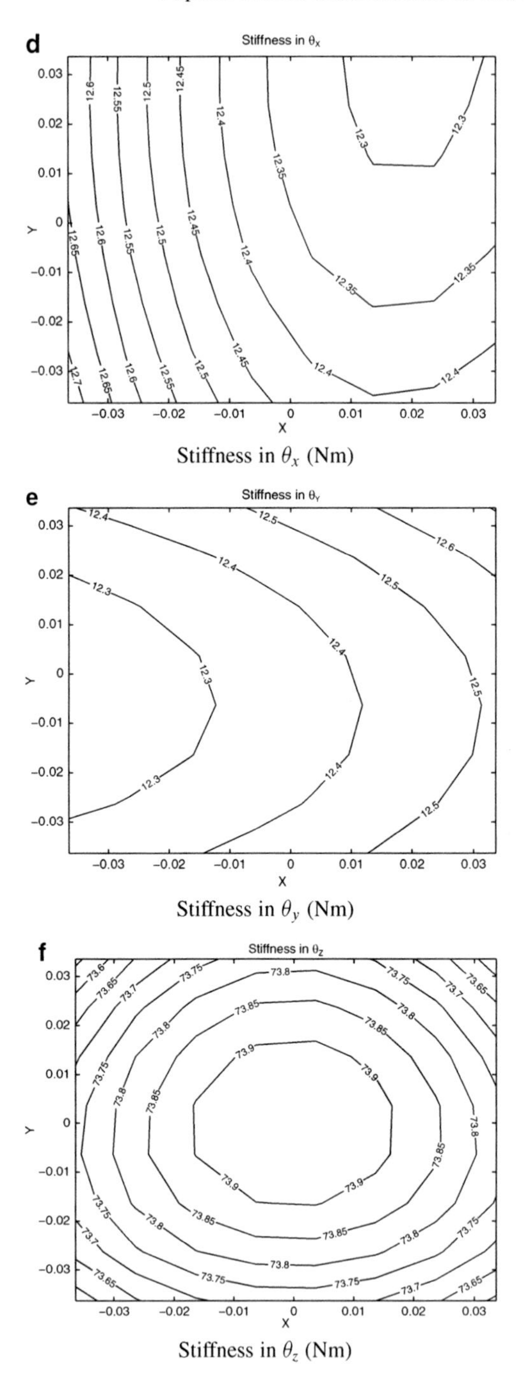

Fig. 6.4 (continued)

platform moves further from the Y-axis. This was to be expected because when the platform moves aside along the X-axis, the projection of the legs on this axis becomes larger, and the mechanism is stiffer in Y. And the same reasoning applies to Fig. 6.4(b) for the stiffness in Y.

In Fig. 6.4d, e, the torsional stiffnesses in θ_x and θ_y are shown, the stiffness is larger when the platform moves further from the Y-axis. However, in the center of the workspace, the K_z is at its minimum, and the stiffness in the Z becomes higher when the platform moves further from the center of the workspace. On the other hand, from Fig. 6.4(f), the stiffness in θ_z is higher near the center of the workspace, which is the best position for supporting torsional loads around Z-axis. All these are in accordance with what would be intuitively expected.

6.3 General Kinematic Model of n Degrees of Freedom Parallel Mechanisms with a Passive Constraining Leg and Revolute Actuators

6.3.1 Geometric Modeling and Lumped Compliance Model

An example of parallel mechanism belonging to the family of mechanisms studied in this chapter is shown in Figs. 6.5 and 6.6. It is a 5-dof parallel mechanism with revolute actuators.

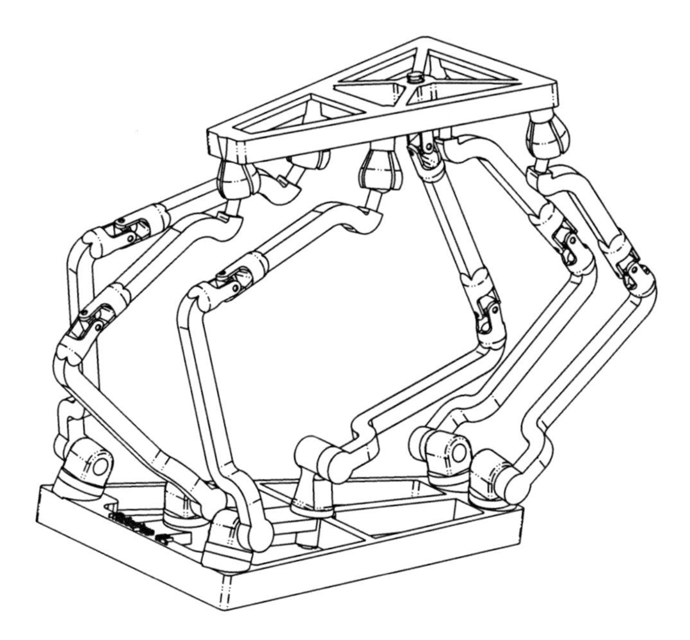

Fig. 6.5 CAD model of the spatial 5-dof parallel mechanism with revolute actuators (Figure by Gabriel Coté)

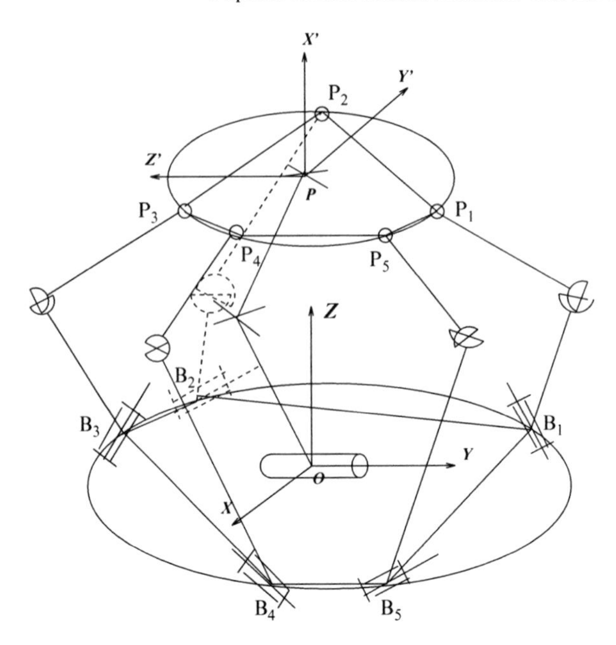

Fig. 6.6 Schematic representation of the spatial 5-dof parallel mechanism with revolute actuators

In order to obtain a simple kinetostatic model, link compliances are lumped at the joints should be considered. In this framework, link bending stiffnesses are replaced by equivalent torsional springs located at virtual joints.

6.3.2 *Inverse Kinematics*

6.3.2.1 Solution for the Case of Mechanisms with Rigid Links

In order to solve the inverse kinematic problem, one must first consider the passive constraining leg as a serial n-dof mechanism whose n Cartesian coordinates are known, which is a well-known problem. Once the solution to the inverse kinematics of this n-dof serial mechanism is found, the complete pose (position and orientation) of the platform can be determined using the direct kinematic equations for this serial mechanism. Figure 6.7 illustrates the configuration of the ith actuated joint of the mechanism with revolute actuators. Point B_i' is defined as the center of the Hooke joint connecting the two moving links of the ith actuated leg. Moreover, the Cartesian coordinates of point B_i' expressed in the fixed coordinate frame are represented as $(b'_{ix}, b'_{iy}, b'_{iz})$ and the position vector of point B_i' in the fixed frame is given by vector $\mathbf{b}_i'$. Since the axis of the fixed revolute joint of the ith actuated leg is assumed to be parallel to the xy plane of the fixed coordinate frame, one can write

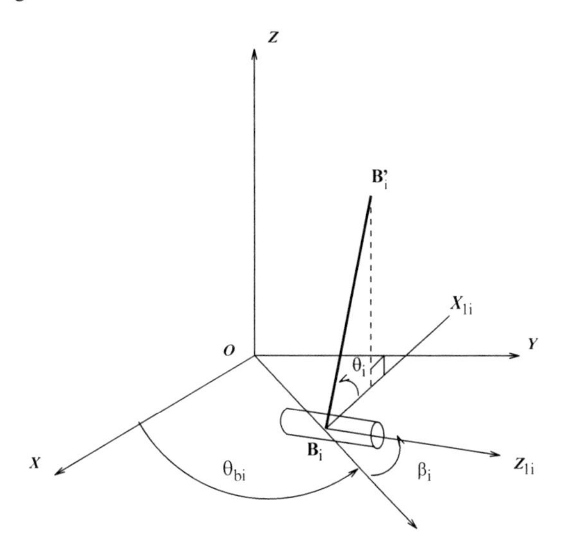

Fig. 6.7 The ith actuated revolute joint

$$b'_{ix} = b_{ix} - l_{i1}\sin(\theta_{bi} + \beta_i)\cos\theta_{i1}, \quad i = 1, \ldots, n, \; n = 3, 4, \text{or } 5, \tag{6.25}$$
$$b'_{iy} = b_{iy} + l_{i1}\cos(\theta_{bi} + \beta_i)\cos\theta_{i1}, \quad i = 1, \ldots, n, \; n = 3, 4, \text{or } 5, \tag{6.26}$$
$$b'_{iz} = b_{iz} + l_{i1}\sin\theta_{i1}, \quad i = 1, \ldots, n, \; n = 3, 4, \text{or } 5, \tag{6.27}$$

where θ_{bi} is the angle between the positive direction of the x-axis of the base coordinate frame and the line connecting points O and B_i and θ_{i1} is the joint variable – rotation angle around the fixed revolute joint – associated with the ith actuated leg, β_i is the angle between the positive direction of the line connecting points O and B_i and the axis of the ith actuated joint. Moreover, l_{i1} is the length of the first link of the ith actuated leg. From the configuration of Fig. 6.7, the relationships between the parameters can be written as

$$(b'_{ix}-x_i)^2+(b'_{iy}-y_i)^2+(b'_{iz}-z_i)^2 = l_{i2}^2, \quad i = 1, \ldots, n, \; n = 3, 4, \text{or } 5, \tag{6.28}$$

where x_i, y_i, z_i are the coordinates of point P_i and l_{i2} is the length of the second link of the ith actuated leg.

Substituting (6.25) – (6.27) into (6.28), one has

$$E_i\cos\theta_{i1} + F_i\sin\theta_{i1} = G_i, \quad i = 1, \ldots, n, \; n = 3, 4, \text{or } 5, \tag{6.29}$$

where

$$E_i = (y_i - b_{iy})\cos(\theta_{bi} + \beta_i) - (x_i - b_{ix})\sin(\theta_{bi} + \beta_i), \tag{6.30}$$
$$F_i = z_i - b_{iz}, \tag{6.31}$$
$$G_i = \frac{(x_i - b_{ix})^2 + (y_i - b_{iy})^2 + (z_i - b_{iz})^2 + l_{i1}^2 - l_{i2}^2}{2l_{i1}} \tag{6.32}$$

and angle θ_{i1} can be obtained by

$$\sin\theta_{i1} = \frac{F_i G_i + K_i E_i \sqrt{H_i}}{E_i^2 + F_i^2}, \quad i = 1, \ldots, n, \ n = 3, 4, \text{or } 5, \tag{6.33}$$

$$\cos\theta_{i1} = \frac{E_i G_i - K_i F_i \sqrt{H_i}}{E_i^2 + F_i^2}, \quad i = 1, \ldots, n, \ n = 3, 4, \text{or } 5, \tag{6.34}$$

where $K_i = \pm 1$ is the branch index of the mechanism associated with the configuration of the ith leg and

$$H_i = E_i^2 + F_i^2 - G_i^2, \quad i = 1, \ldots, n, \ n = 3, 4, \text{or } 5. \tag{6.35}$$

Finally, the solution of the inverse kinematic problem is completed by performing

$$\theta_{i1} = \text{atan2}[\sin\theta_{i1}, \cos\theta_{i1}], \quad i = 1, \ldots, n, \ n = 3, 4, \text{or } 5. \tag{6.36}$$

Meanwhile, referring to Fig. 6.7, the vector of leg length can be written as

$$\mathbf{b}_i' = \mathbf{b}_i + l_{i1}\mathbf{Q}_{ti1}\mathbf{d}_i, \quad i = 1, \ldots, n, \ n = 3, 4, \text{or } 5 \tag{6.37}$$

with

$$\mathbf{Q}_{ti1} = \begin{bmatrix} \cos(\theta_{bi} + \beta_i) & -\sin(\theta_{bi} + \beta_i) & 0 \\ \sin(\theta_{bi} + \beta_i) & \cos(\theta_{bi} + \beta_i) & 0 \\ 0 & 0 & 1 \end{bmatrix}, \quad i = 1, \ldots, n, \ n = 3, 4, \text{or } 5 \tag{6.38}$$

and

$$\mathbf{d}_{i1} = \begin{bmatrix} 0 \\ \cos\theta_{i1} \\ \sin\theta_{i1} \end{bmatrix}, \quad i = 1, \ldots, n, \ n = 3, 4, \text{or } 5 \tag{6.39}$$

assuming that the distance between points P_i and B_i' is noted l_{i2}, then one has

$$l_{i2}^2 = (\mathbf{p}_i - \mathbf{b}_i)^{\mathrm{T}}(\mathbf{p}_i - \mathbf{b}_i), \quad i = 1, \ldots, n, \quad n = 3, 4, \text{or } 5. \tag{6.40}$$

6.3.2.2 Solutions for the Mechanisms with Flexible Links

In order to uniquely describe the architecture of a kinematic chain, i.e., the relative location and orientation of its neighboring pair axes, the Denavit–Hartenberg notation is used to define the nominal geometry of each of the serial kinematic chains of the parallel mechanism. A coordinate frame F_i is defined with the origin O_i and axes X_i, Y_i, Z_i, this frame is attached to the $(i-1)$th link. Figure 6.8 represents one of the identical kinematic chains for the n-dof parallel mechanism discussed above. Joint 2 is a virtual joint used to model the compliance of the driven link.

Fig. 6.8 One of the identical kinematic chains with flexible links

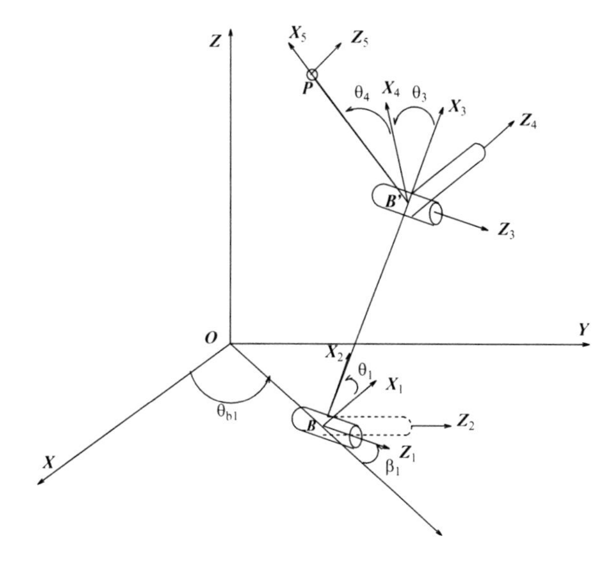

From Fig. 6.8, one has $\theta_{i2} = 0$, when there is no deflection. Angles θ_{i3} and θ_{i4} can be obtained by writing the coordinates of point P_i in Frame 3 as

$$x_{i3} = l_{i2} \cos\theta_{i4} \cos\theta_{i3}, \quad i = 1, \ldots, n, \; n = 3, 4, \text{ or } 5, \tag{6.41}$$

$$y_{i3} = l_{i2} \cos\theta_{i4} \sin\theta_{i3}, \quad i = 1, \ldots, n, \; n = 3, 4, \text{ or } 5, \tag{6.42}$$

$$z_{i3} = l_{i2} \sin\theta_{i4}, \quad i = 1, \ldots, n, \; n = 3, 4, \text{ or } 5, \tag{6.43}$$

and

$$[\mathbf{p}]_3 = \mathbf{Q}_{i2}^T \mathbf{Q}_{i1}^T \mathbf{Q}_{i0}^T [\mathbf{p}_i - \mathbf{b}'_i], \quad i = 1, \ldots, n, \; n = 3, 4, \text{ or } 5. \tag{6.44}$$

then, combining (6.41) – (6.43) and (6.44), one can find θ_{i3} and θ_{i4} easily.

From Fig. 6.8, one can express the position of point B'_i as

$$\mathbf{b}'_i = \mathbf{b}_i + \mathbf{Q}_{i0}\mathbf{a}_{i1} + \mathbf{Q}_{i0}\mathbf{Q}_{i1}\mathbf{a}_{i2}, \quad i = 1, \ldots, n, \; n = 3, 4, \text{ or } 5, \tag{6.45}$$

where $\mathbf{Q}_{i0}$, $\mathbf{a}_{i1}$, $\mathbf{a}_{i2}$ and $\mathbf{Q}_{i1}$ can be expressed as

$$\mathbf{a}_{i1} = \begin{bmatrix} 0 \\ 0 \\ 0 \end{bmatrix}, \; \mathbf{a}_{i2} = \begin{bmatrix} l_{i1} \cos\theta_{i2} \\ l_{i1} \sin\theta_{i2} \\ 0 \end{bmatrix}, \; \mathbf{Q}_{i1} = \begin{bmatrix} \cos\theta_{i1} & 0 & \sin\theta_{i1} \\ \sin\theta_{i1} & 0 & -\cos\theta_{i1} \\ 0 & 1 & 0 \end{bmatrix}, \tag{6.46}$$

$$\mathbf{Q}_{i0} = \begin{bmatrix} -\sin(\theta_{bi} + \beta_i) & 0 & \cos(\theta_{bi} + \beta_i) \\ \cos(\theta_{bi} + \beta_i) & 0 & \sin(\theta_{bi} + \beta_i) \\ 0 & 1 & 0 \end{bmatrix}. \tag{6.47}$$

6.3.3 *Jacobian Matrices*

6.3.3.1 Rigid Mechanisms

The parallel mechanisms studied here comprise two main components, namely, the passive constraining leg – which can be thought of as a serial mechanism – and the actuated legs acting in parallel.

Considering the passive constraining leg, one can write

$$\mathbf{J}_{n+1}\dot{\theta}_{n+1} = \mathbf{t}, \; n = 3, 4, \text{ or } 5, \tag{6.48}$$

where $\mathbf{J}_{n+1}$ consists of $\mathbf{e}_i$ and $\mathbf{r}_i$, $\mathbf{t} = \begin{bmatrix} \omega^{\mathrm{T}} \; \dot{\mathbf{p}}^{\mathrm{T}} \end{bmatrix}^{\mathrm{T}}$ is the twist of the platform, ω is the angular velocity of the platform, and $\dot{\theta}_{n+1} = \begin{bmatrix} \dot{\theta}_{n+1,1} \cdots \dot{\theta}_{n+1,n} \end{bmatrix}^{\mathrm{T}}$, ($n =$ 3, 4, or 5) is the joint velocity vector associated with the passive constraining leg. Matrix $\mathbf{J}_{n+1}$ is the Jacobian matrix of the passive constraining leg which is taken as a serial n-dof mechanism.

6.3.3.2 Compliant Model

If the compliances of the links and joints are included, $(6-n)$ virtual joints will then be added to the passive constraining leg in order to account for the compliance of the links [62]. Hence, the Jacobian matrix of the passive constraining leg becomes

$$\mathbf{J}'_{n+1}\dot{\theta}'_{n+1} = \mathbf{t}, \quad n = 3, 4, \text{ or } 5, \tag{6.49}$$

where

$$\dot{\theta}'_{n+1} = \begin{bmatrix} \dot{\theta}_{n+1,1} \cdots \dot{\theta}_{n+1,6} \end{bmatrix}^{\mathrm{T}}, \quad n = 3, 4, \text{ or } 5. \tag{6.50}$$

6.3.3.3 Global Velocity Equations

1. Rigid Model:
   Now, considering the parallel component of the mechanism, the parallel Jacobian matrix can be obtained by differentiating (6.37), (6.39), and (6.40) with respect to time. One has

$$\dot{\mathbf{b}}'_i = l_{i1}\mathbf{Q}_{ti1}\dot{\mathbf{d}}_i, \quad i = 1, \ldots, n, \; n = 3, 4, \text{ or } 5, \tag{6.51}$$

$$\dot{\mathbf{d}}_{i1} = \begin{bmatrix} 0 \\ -\sin\theta_{i1} \\ \cos\theta_{i1} \end{bmatrix} \dot{\theta}_{i1}, \quad i = 1, \ldots, n, \; n = 3, 4, \text{ or } 5, \tag{6.52}$$

$$(\mathbf{p}_i - \mathbf{b}'_i)^{\mathrm{T}}\dot{\mathbf{b}}'_i - (\mathbf{p}_i - \mathbf{b}'_i)^{\mathrm{T}}\dot{\mathbf{p}}_i = 0, \quad i = 1, \ldots, n, \; n = 3, 4, \text{ or } 5. \tag{6.53}$$

One obtains

$$\dot{\mathbf{p}}_i = \dot{\mathbf{p}} + \dot{\mathbf{Q}}\mathbf{r'}_i, \quad i = 1, \ldots, n, \ n = 3, 4, \text{or } 5 \tag{6.54}$$

assuming

$$\mathbf{e}_i = l_{i1}\mathbf{Q}_{ti1}\begin{bmatrix} 0 \\ -\sin\theta_{i1} \\ \cos\theta_{i1} \end{bmatrix}, \quad i = 1, \ldots, n, \ n = 3, 4, \text{or } 5 \tag{6.55}$$

then

$$\dot{\mathbf{b}}'_i = \mathbf{e}_i\dot{\theta}_{i1}, \quad i = 1, \ldots, n, \ n = 3, 4, \text{or } 5 \tag{6.56}$$

therefore (6.53) can be rewritten as (for $i = 1, \ldots, n, \quad n = 3, 4, \text{or } 5$)

$$(\mathbf{p}_i - \mathbf{b}'_i)^{\mathrm{T}}\mathbf{e}_i\dot{\theta}_{i1} = (\mathbf{p}_i - \mathbf{b}'_i)^{\mathrm{T}}\dot{\mathbf{p}} + [(\mathbf{Q}\mathbf{r'}_i) \times (\mathbf{p}_i - \mathbf{b}'_i)]^{\mathrm{T}}\omega. \tag{6.57}$$

Hence, one has the velocity equation as

$$\mathbf{A}\mathbf{t} = \mathbf{B}\dot{\theta}, \tag{6.58}$$

where vector $\dot{\theta}$ and $\mathbf{t}$ are defined as

$$\dot{\theta} = \begin{bmatrix} \dot{\theta}_1 \cdots \dot{\theta}_n \end{bmatrix}^{\mathrm{T}}, \quad n = 3, 4, \text{or } 5, \tag{6.59}$$

$$\mathbf{t} = \begin{bmatrix} \omega_1 \ \omega_2 \ \omega_3 \ \dot{x} \ \dot{y} \ \dot{z} \end{bmatrix}^{\mathrm{T}}, \tag{6.60}$$

vector $\boldsymbol{\omega}$ is the angular velocity of the platform, and

$$\mathbf{A} = \begin{bmatrix} \mathbf{a}_1^{\mathrm{T}} \\ \mathbf{a}_2^{\mathrm{T}} \\ \vdots \\ \mathbf{a}_n^{\mathrm{T}} \end{bmatrix}, \quad \mathbf{B} = \mathrm{diag}[(\mathbf{p}_1 - \mathbf{b}'_1)^{\mathrm{T}}\mathbf{e}_1, \ldots, (\mathbf{p}_n - \mathbf{b}'_n)^{\mathrm{T}}\mathbf{e}_n], \tag{6.61}$$

where $\mathbf{a}_i$ is a six-dimensional vector, which can be expressed as

$$\mathbf{a}_i = \begin{bmatrix} (\mathbf{Q}\mathbf{r'}_i) \times (\mathbf{p}_i - \mathbf{b}'_i) \\ (\mathbf{p}_i - \mathbf{b}'_i) \end{bmatrix}, \quad i = 1, \ldots, n, \ n = 3, 4, \text{or } 5. \tag{6.62}$$

2. Compliant Model:
   Differentiating (6.45) and (6.46) with respect to time, one has

$$\dot{\mathbf{b}}'_i = \mathbf{Q}_{i0}\dot{\mathbf{Q}}_{i1}\mathbf{a}_{i2} + \mathbf{Q}_{i0}\mathbf{Q}_{i1}\dot{\mathbf{a}}_{i2}, \quad i = 1, \ldots, n, \ n = 3, 4, \text{or } 5, \tag{6.63}$$

$$\dot{\mathbf{a}}_{i2} = \begin{bmatrix} -l_{i1}\sin\theta_{i2} \\ l_{i1}\cos\theta_{i2} \\ 0 \end{bmatrix} \dot{\theta}_{i2}, \quad i = 1, \ldots, n, \tag{6.64}$$

$$\dot{\mathbf{Q}}_{i1} = \begin{bmatrix} -\sin\theta_{i1} & 0 & \cos\theta_{i1} \\ \cos\theta_{i1} & 0 & \sin\theta_{i1} \\ 0 & 0 & 0 \end{bmatrix} \dot{\theta}_{i1}, \quad i = 1, \ldots, n, \; n = 3, 4, \text{or } 5. \tag{6.65}$$

For $i = 1, \ldots, n, \quad n = 3, 4, \text{or } 5$, assuming

$$\mathbf{d}_{i1} = \mathbf{Q}_{i0} \begin{bmatrix} -\sin\theta_{i1} & 0 & \cos\theta_{i1} \\ \cos\theta_{i1} & 0 & \sin\theta_{i1} \\ 0 & 0 & 0 \end{bmatrix} \mathbf{a}_{i2}, \; \mathbf{d}_{i2} = \mathbf{Q}_{i0}\mathbf{Q}_{i1} \begin{bmatrix} -l_{i1}\sin\theta_{i2} \\ l_{i1}\cos\theta_{i2} \\ 0 \end{bmatrix} \tag{6.66}$$

then one has

$$\dot{\mathbf{b}}_i' = \mathbf{d}_{i1}\dot{\theta}_{i1} + \mathbf{d}_{i2}\dot{\theta}_{i2}, \quad i = 1, \ldots, n, \; n = 3, 4, \text{or } 5. \tag{6.67}$$

Differentiating (6.40) with respect to time, one obtains (6.53), and following a derivation similar to the one presented above for the mechanism with rigid links, for $i = 1, \ldots, n, \quad n = 3, 4, \text{or } 5$, one obtains

$$(\mathbf{p}_i - \mathbf{b}_i')^{\mathrm{T}}(\mathbf{d}_{i1}\dot{\theta}_{i1} + \mathbf{d}_{i2}\dot{\theta}_{i2}) = (\mathbf{p}_i - \mathbf{b}_i')^{\mathrm{T}}\dot{\mathbf{p}} + [(\mathbf{Q}\mathbf{r}'_i) \times (\mathbf{p}_i - \mathbf{b}_i')]^{\mathrm{T}}\omega. \tag{6.68}$$

Hence one has the velocity equation as

$$\mathbf{At} = \mathbf{B}_1\dot{\theta}_1 + \mathbf{B}_2\dot{\theta}_2, \tag{6.69}$$

where vectors $\dot{\theta}_1$ and $\dot{\theta}_2$ are defined as

$$\dot{\theta}_1 = \begin{bmatrix} \dot{\theta}_{11} \cdots \dot{\theta}_{n1} \end{bmatrix}^{\mathrm{T}}, \quad n = 3, 4, \text{or } 5, \tag{6.70}$$

$$\dot{\theta}_2 = \begin{bmatrix} \dot{\theta}_{12} \cdots \dot{\theta}_{n2} \end{bmatrix}^{\mathrm{T}}, \quad n = 3, 4, \text{or } 5, \tag{6.71}$$

matrices $\mathbf{A}$, $\mathbf{B}_1$ and $\mathbf{B}_2$ are given as

$$\mathbf{A} = \begin{bmatrix} \mathbf{a}_1 \; \mathbf{a}_2 \; \mathbf{a}_3 \; \mathbf{a}_4 \; \mathbf{a}_5 \; \mathbf{a}_6 \end{bmatrix}^{\mathrm{T}} \tag{6.72}$$

$$\mathbf{B}_1 = \mathrm{diag}[b_{11}, \ldots, b_{n,2n-1}], \quad n = 3, 4, \text{or } 5, \tag{6.73}$$

$$\mathbf{B}_2 = \mathrm{diag}[b_{12}, \ldots, b_{n,2n}], \quad n = 3, 4, \text{or } 5, \tag{6.74}$$

where $\mathbf{a}_i$ is a six-dimensional vector and $\mathbf{b}_{i,2i-1}$, $\mathbf{b}_{i,2i}$ are the diagonal items of $\mathbf{B}_1$ and $\mathbf{B}_2$, respectively. They can be expressed as

$$\mathbf{a}_i = \begin{bmatrix} (\mathbf{Q}\mathbf{r}'_i) \times (\mathbf{p}_i - \mathbf{b}'_i) \\ (\mathbf{p}_i - \mathbf{b}'_i) \end{bmatrix}, \quad i = 1, \ldots, n, \; n = 3, 4, \text{ or } 5, \tag{6.75}$$

$$b_{i,2i-1} = (\mathbf{p}_i - \mathbf{b}'_i)^{\mathrm{T}} \mathbf{d}_{i,1}, \quad i = 1, \ldots, n, \; n = 3, 4, \text{ or } 5, \tag{6.76}$$

$$b_{i,2i} = (\mathbf{p}_i - \mathbf{b}'_i)^{\mathrm{T}} \mathbf{d}_{i,2}, \quad i = 1, \ldots, n, \; n = 3, 4, \text{ or } 5. \tag{6.77}$$

6.3.4 Kinetostatic Model for the Mechanism with Rigid Links

According to the principle of virtual work, one can finally obtain the Cartesian compliance matrix with the same approach as in Chap. 5.

$$\mathbf{C}_{\mathrm{c}} = \mathbf{J}_{n+1} (\mathbf{A}\mathbf{J}_{n+1})^{-1} \mathbf{B}\mathbf{C}\mathbf{B}^{\mathrm{T}} (\mathbf{A}\mathbf{J}_{n+1})^{-\mathrm{T}} \mathbf{J}_{n+1}^{\mathrm{T}} \tag{6.78}$$

with

$$\Delta \mathbf{c} = \mathbf{C}_{\mathrm{c}} \mathbf{w}, \tag{6.79}$$

where $\mathbf{C}_{\mathrm{c}}$ is a symmetric positive semidefinite (6×6) matrix, as expected.

6.3.5 Kinetostatic Model for the Mechanism with Flexible Links

Again, based on the principle of virtual work, one can write

$$\mathbf{w}^{\mathrm{T}} \mathbf{t} = \tau_{n+1}^{\mathrm{T}} \dot{\theta}'_{n+1} + \tau_1^{\mathrm{T}} \dot{\theta}_1 + \tau_2^{\mathrm{T}} \dot{\theta}_2, \tag{6.80}$$

where τ_1 and τ_2 correspond to a partition of vector τ, in components associated with $\dot{\theta}_1$ and $\dot{\theta}_2$, respectively, i.e., the first and second joint of each leg. τ is the vector of actuator forces, $\dot{\theta}$ is the vector of actuator velocities (actuated joints and joints with virtual springs), and τ_{n+1} is the vector of joint torques in the passive constraining leg. This vector is defined as follows, where $\mathbf{K}_{n+1}$ is the stiffness matrix of the passive constraining leg,

$$\tau_{n+1} = \mathbf{K}_{n+1} \Delta \theta'_{n+1}, \tag{6.81}$$

$$\tau_1 = \mathbf{K}_{j1} \Delta \theta_1, \tag{6.82}$$

$$\tau_2 = \mathbf{K}_{j2} \Delta \theta_2, \tag{6.83}$$

$$\mathbf{K}_{j1} = \operatorname{diag}[k_{11}, \ldots, k_{n1}], \tag{6.84}$$

$$\mathbf{K}_{j2} = \operatorname{diag}[k_{12}, \ldots, k_{n2}]. \tag{6.85}$$

Matrix $\mathbf{K}_{n+1}$ is a diagonal 6×6 matrix in which the ith diagonal entry is zero if it is associated with a real joint or it is equal to k_i if it is associated with a virtual joint, where k_i is the stiffness of the virtual spring located at the ith joint. $k_{11}, \ldots, k_{n1}$ are the compound stiffnesses of actuators and first links stiffnesses while $k_{12}, \ldots, k_{n2}$ are the first links stiffnesses. One can rewrite (6.69) as

$$\dot{\theta}_1 = \mathbf{B}_1^{-1}\mathbf{At} - \mathbf{B}_1^{-1}\mathbf{B}_2\dot{\theta}_2. \tag{6.86}$$

Substituting (6.86) and (6.49) into (6.80), one can obtain

$$\mathbf{w}^{\mathrm{T}}\mathbf{J}'_{n+1}\dot{\theta}'_{n+1} = \tau_{n+1}^{\mathrm{T}}\dot{\theta}'_{n+1} + \tau_2^{\mathrm{T}}\dot{\theta}_2 + \tau_1^{\mathrm{T}}\mathbf{B}_1^{-1}\mathbf{A}\mathbf{J}'_{n+1}\dot{\theta}'_{n+1} - \tau_1^{\mathrm{T}}\mathbf{B}_1^{-1}\mathbf{B}_2\dot{\theta}_2. \tag{6.87}$$

Since there are 11 degrees of freedom in the compliant mechanism, this equation must be satisfied for any value of $\dot{\theta}'_{n+1}$ and $\dot{\theta}_2$. Therefore, one can equate the coefficients of the terms in $\dot{\theta}'_{n+1}$ and the terms in $\dot{\theta}_2$, hence one can obtain

$$(\mathbf{J}'_{n+1})^{\mathrm{T}}\mathbf{w} = \tau_{n+1} + (\mathbf{J}'_{n+1})^{\mathrm{T}}\mathbf{A}^{\mathrm{T}}\mathbf{B}_1^{-\mathrm{T}}\tau_1, \tag{6.88}$$

$$\tau_2 = \mathbf{B}_2^{\mathrm{T}}\mathbf{B}_1^{-\mathrm{T}}\tau_1. \tag{6.89}$$

Substituting (6.81), (6.82), and (6.83) into (6.88) and (6.89), one obtains

$$(\mathbf{J}'_{n+1})^{\mathrm{T}}\mathbf{w} = \mathbf{K}_{n+1}\Delta\theta'_{n+1} + (\mathbf{J}'_{n+1})^{\mathrm{T}}\mathbf{A}^{\mathrm{T}}\mathbf{B}_1^{-\mathrm{T}}\mathbf{K}_{j1}\Delta\theta_1, \tag{6.90}$$

$$\Delta\theta_2 = \mathbf{K}_{j2}^{-1}\mathbf{B}_2^{\mathrm{T}}\mathbf{B}_1^{-\mathrm{T}}\mathbf{K}_{j1}\Delta\theta_1. \tag{6.91}$$

Substituting (6.91) into (6.69), one obtains

$$\mathbf{At} = \mathbf{W}\dot{\theta}_1, \tag{6.92}$$

where

$$\mathbf{W} = \mathbf{B}_1 + \mathbf{B}_2\mathbf{K}_{j2}^{-1}\mathbf{B}_2^{\mathrm{T}}\mathbf{B}_1^{-\mathrm{T}}\mathbf{K}_{j1}. \tag{6.93}$$

Substituting (6.92) into (6.90), one obtains

$$(\mathbf{J}'_{n+1})^{\mathrm{T}}\mathbf{w} = \mathbf{K}_{n+1}(\mathbf{J}'_{n+1})^{-1}\Delta\mathbf{c} + (\mathbf{J}'_{n+1})^{\mathrm{T}}\mathbf{A}^{\mathrm{T}}\mathbf{B}_1^{-\mathrm{T}}\mathbf{K}_{j1}\mathbf{W}^{-1}\mathbf{A}\Delta\mathbf{c}, \tag{6.94}$$

i.e.,

$$\mathbf{w} = ((\mathbf{J}'_{n+1})^{-\mathrm{T}}\mathbf{K}_{n+1}(\mathbf{J}'_{n+1})^{-1} + \mathbf{A}^{\mathrm{T}}\mathbf{B}_1^{-\mathrm{T}}\mathbf{K}_{j1}\mathbf{W}^{-1}\mathbf{A})\Delta\mathbf{c}, \tag{6.95}$$

which is in the form

$$\mathbf{w} = \mathbf{K}\Delta\mathbf{c}, \tag{6.96}$$

where $\mathbf{K}$ is the stiffness matrix, which is equal to

$$\mathbf{K} = [(\mathbf{J}'_{n+1})^{-\mathrm{T}}\mathbf{K}_{n+1}(\mathbf{J}'_{n+1})^{-1} + \mathbf{A}^{\mathrm{T}}\mathbf{B}_1^{-\mathrm{T}}\mathbf{K}_{j1}\mathbf{W}^{-1}\mathbf{A}]. \tag{6.97}$$

Matrix $\mathbf{K}$ is a symmetric (6×6) positive semidefinite matrix, as expected. Matrix $\mathbf{K}$ will be of full rank in nonsingular configurations. Indeed, the sum of the two terms in (6.97) will span the complete space of constraint wrenches.

6.3.6 Examples

6.3.6.1 5-dof Parallel Mechanism

This mechanism is illustrated is Fig. 6.5, the compliance matrix for the mechanism with rigid links can be written as

$$\mathbf{C}_c = \mathbf{J}_6(\mathbf{A}\mathbf{J}_6)^{-1}\mathbf{B}\mathbf{C}\mathbf{B}^{\mathrm{T}}(\mathbf{A}\mathbf{J}_6)^{-\mathrm{T}}\mathbf{J}_6^{\mathrm{T}}, \tag{6.98}$$

where

$$\mathbf{C} = \mathrm{diag}[c_1, c_2, c_3, c_4, c_5] \tag{6.99}$$

with c_1, c_2, c_3, c_4 and c_5 the compliances of the actuators and $\mathbf{J}_6$ is the Jacobian matrix of the passive constraining leg in this 5-dof case. Matrices $\mathbf{A}$ and $\mathbf{B}$ are the Jacobian matrices of the structure without the passive constraining leg.

Similarly, the stiffness matrix for the mechanism with flexible links can be written as

$$\mathbf{K} = (\mathbf{J}_6')^{-\mathrm{T}}\mathbf{K}_6(\mathbf{J}_6')^{-1} + \mathbf{A}^{\mathrm{T}}\mathbf{B}_1^{-\mathrm{T}}\mathbf{K}_{j1}\mathbf{W}^{-1}\mathbf{A}, \tag{6.100}$$

where

$$\mathbf{K}_6 = \mathrm{diag}[0, k_{62}, 0, 0, 0, 0], \tag{6.101}$$
$$\mathbf{K}_{j1} = \mathrm{diag}[k_{11}, k_{21}, k_{31}, k_{41}, k_{51}], \tag{6.102}$$
$$\mathbf{K}_{j2} = \mathrm{diag}[k_{12}, k_{22}, k_{32}, k_{42}, k_{52}], \tag{6.103}$$

where k_{62} is the stiffness of the virtual joint of the passive constraining leg, and $\mathbf{J}_6'$ is the Jacobian matrix of the passive constraining leg in this 5-dof case. Matrices $\mathbf{A}$ and $\mathbf{B}$ are the Jacobian matrices of the structure without the passive constraining leg.

The comparison between the parallel mechanism with rigid links (without virtual joints) and the parallel mechanism with flexible links (with virtual joints) is given in Table 6.2.

6.3.6.2 4-dof Parallel Mechanism

This mechanism is illustrated is Fig. 6.9, the compliance matrix for the mechanism with rigid links will be

$$\mathbf{C}_c = \mathbf{J}_5(\mathbf{A}\mathbf{J}_5)^{-1}\mathbf{B}\mathbf{C}\mathbf{B}^{\mathrm{T}}(\mathbf{A}\mathbf{J}_5)^{-\mathrm{T}}\mathbf{J}_5^{\mathrm{T}} \tag{6.104}$$

Table 6.2 Comparison of the 5-dof mechanism compliance between the mechanism with flexible links and the mechanism with rigid links

K_{actuator}	K_{link}	κ_{θ_x}	κ_{θ_y}	κ_{θ_z}	κ_x	κ_y	κ_z
1,000	1,000	0.255808	0.478997	0.766154	0.00479741	0.0116413	0.00169207
1,000	$10^1 K_a$	0.137552	0.257339	0.412528	0.00257412	0.00625667	0.000910745
1,000	$10^2 K_a$	0.125726	0.235173	0.377165	0.0023518	0.0057182	0.000832613
1,000	$10^3 K_a$	0.124543	0.232956	0.373629	0.00232956	0.00566435	0.000824799
1,000	$10^4 K_a$	0.124425	0.232734	0.373275	0.00232734	0.00565897	0.000824018
1,000	$10^5 K_a$	0.124413	0.232712	0.37324	0.00232712	0.00565843	0.00082394
1,000	$10^6 K_a$	0.124412	0.23271	0.373236	0.00232709	0.00565838	0.000823932
1,000	$10^7 K_a$	0.124412	0.23271	0.373236	0.00232709	0.00565837	0.000823931
1,000	Rigid	0.124412	0.23271	0.373236	0.00232709	0.00565837	0.000823931

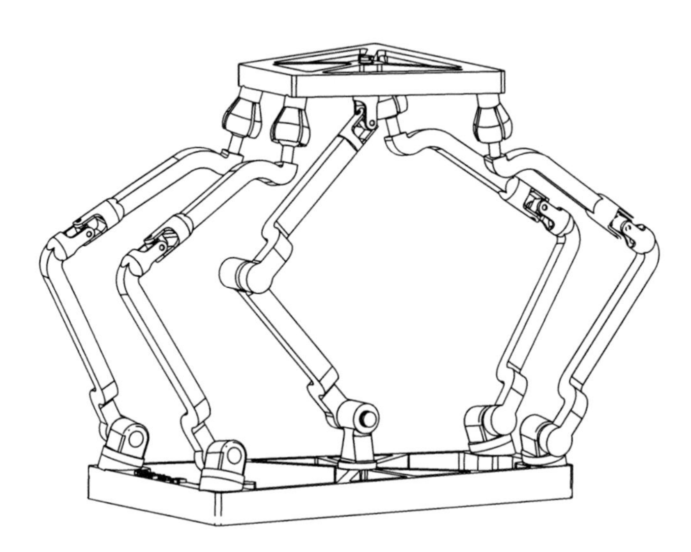

Fig. 6.9 CAD model of the spatial 4-dof parallel mechanism with revolute actuators (Figure by Gabriel Coté)

where

$$\mathbf{C} = \text{diag}[c_1, c_2, c_3, c_4] \tag{6.105}$$

with c_1, c_2, c_3, and c_4 are the compliances of the actuators and $\mathbf{J}_5$ is the Jacobian matrix of the constraining leg in this 4-dof case. Matrices $\mathbf{A}$ and $\mathbf{B}$ are the Jacobian matrices of the structure without the passive constraining leg.

Similarly, the stiffness matrix for the mechanism with flexible links can be written as

$$\mathbf{K} = (\mathbf{J}'_5)^{-\mathrm{T}}\mathbf{K}_5(\mathbf{J}'_5)^{-1} + \mathbf{A}^{\mathrm{T}}\mathbf{B}_1^{-\mathrm{T}}\mathbf{K}_{j1}\mathbf{W}^{-1}\mathbf{A}, \tag{6.106}$$

Table 6.3 Comparison of the 4-dof mechanism compliance between the mechanism with flexible links and the mechanism with rigid links

K_{actuator}	K_{link}	κ_{θ_x}	κ_{θ_y}	κ_{θ_z}	κ_x	κ_y	κ_z
1,000	1,000	3.46122	0.138363	1.5×10^{-3}	0.0146575	1.32691×10^{-3}	0.000271569
1,000	$10^1 K_a$	1.49876	0.0703633	1.5×10^{-4}	0.00717397	1.32691×10^{-4}	0.000106446
1,000	$10^2 K_a$	1.30251	0.0635633	1.5×10^{-5}	0.00642561	1.32691×10^{-5}	0.0000899335
1,000	$10^3 K_a$	1.28289	0.0628833	1.5×10^{-6}	0.00635078	1.32691×10^{-6}	0.0000882822
1,000	$10^4 K_a$	1.28093	0.0628153	1.5×10^{-7}	0.0063433	1.32691×10^{-7}	0.0000881171
1,000	$10^5 K_a$	1.28073	0.0628085	1.5×10^{-8}	0.00634255	1.32691×10^{-8}	0.0000881006
1,000	$10^6 K_a$	1.28071	0.0628079	1.5×10^{-9}	0.00634247	1.32691×10^{-9}	0.000088099
1,000	$10^7 K_a$	1.28071	0.0628078	1.5×10^{-10}	0.00634246	1.32691×10^{-10}	0.0000880988
1,000	Rigid	1.28071	0.0628078	0.0	0.00634246	0.0	0.0000880988

where

$$\mathbf{K}_5 = \text{diag}[0, k_{52}, 0, k_{54}, 0, 0], \tag{6.107}$$

$$\mathbf{K}_{j1} = \text{diag}[k_{11}, k_{21}, k_{31}, k_{41}], \tag{6.108}$$

$$\mathbf{K}_{j2} = \text{diag}[k_{12}, k_{22}, k_{32}, k_{42}], \tag{6.109}$$

where k_{52} and k_{54} are the stiffnesses of the virtual joints of the passive constraining leg, $\mathbf{J}'_5$ is the Jacobian matrix of the passive constraining leg in this 4-dof case, while $\mathbf{A}$ and $\mathbf{B}_1$, $\mathbf{B}_2$ are the Jacobian matrices of the structure without the passive constraining leg.

The comparison between the parallel mechanism with rigid links (without virtual joints) and the parallel mechanism with flexible links (with virtual joints) is given in Table 6.3. Again, the effect of link flexibility is clearly demonstrated.

6.3.6.3 3-dof Parallel Mechanism

This mechanism is illustrated is Fig. 6.10, the compliance matrix for the rigid mechanism can be written as

$$\mathbf{C}_c = \mathbf{J}_4(\mathbf{A}\mathbf{J}_4)^{-1}\mathbf{B}\mathbf{C}\mathbf{B}^{\mathrm{T}}(\mathbf{A}\mathbf{J}_4)^{-\mathrm{T}}\mathbf{J}_4^{\mathrm{T}}, \tag{6.110}$$

where $\mathbf{C}_c = \text{diag}[c_1, c_2, c_3]$, with c_1, c_2 and c_3 are the compliances of the actuators and $\mathbf{J}_4$ is the Jacobian matrix of the passive constraining leg in this 3-dof case. $\mathbf{A}$ and $\mathbf{B}$ are the Jacobian matrices of the structure without the passive constraining leg.

Similarly, the stiffness matrix for the mechanism with flexible links will be written as

$$\mathbf{K} = [(\mathbf{J}'_4)^{-\mathrm{T}}\mathbf{K}_4(\mathbf{J}'_4)^{-1} + \mathbf{A}^{\mathrm{T}}\mathbf{B}_1^{-\mathrm{T}}\mathbf{K}_{j1}\mathbf{W}^{-1}\mathbf{A}] \tag{6.111}$$

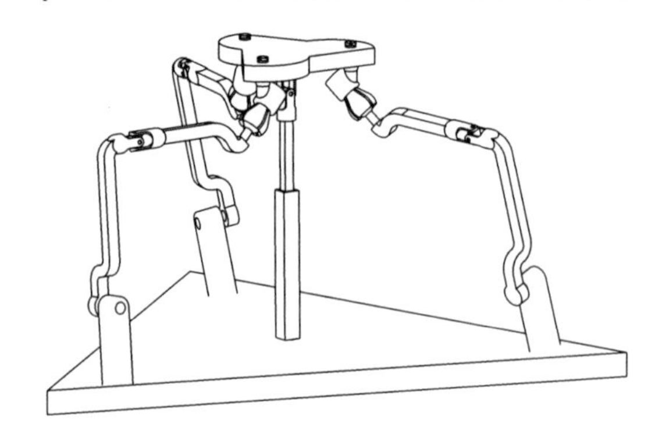

Fig. 6.10 CAD model of the spatial 3-dof parallel mechanism with revolute actuators (Figure by Gabriel Coté)

Table 6.4 Comparison of the 3-dof mechanism compliance between the mechanism with flexible links and the mechanism with rigid links

K_{actuator}	K_{link}	κ_{θ_x}	κ_{θ_y}	κ_{θ_z}	κ_x	κ_y	κ_z
1,000	1,000	0.09937	0.09937	10^{-3}	1.06152×10^{-3}	1.06152×10^{-3}	0.000186579
1,000	$10^1 K_a$	0.02904	0.02904	10^{-4}	1.06152×10^{-4}	1.06152×10^{-4}	0.0000975989
1,000	$10^2 K_a$	0.02201	0.02201	10^{-5}	1.06152×10^{-5}	1.06152×10^{-5}	0.0000887009
1,000	$10^3 K_a$	0.02131	0.02131	10^{-6}	1.06152×10^{-6}	1.06152×10^{-6}	0.0000878111
1,000	$10^4 K_a$	0.02123	0.02123	10^{-7}	1.06152×10^{-7}	1.06152×10^{-7}	0.0000877221
1,000	$10^5 K_a$	0.02123	0.02123	10^{-8}	1.06152×10^{-8}	1.06152×10^{-8}	0.0000877132
1,000	$10^6 K_a$	0.02123	0.02123	10^{-9}	1.06152×10^{-9}	1.06152×10^{-9}	0.0000877123
1,000	$10^7 K_a$	0.02123	0.02123	10^{-10}	1.06152×10^{-10}	1.06152×10^{-10}	0.0000877122
1,000	Rigid	0.02123	0.02123	0.0	0.0	0.0	0.0000877122

where **W** is defined in (6.93) and

$$\mathbf{K}_4 = \text{diag}[k_{41}, k_{42}, k_{43}, 0, 0, 0], \tag{6.112}$$

$$\mathbf{K}_{j1} = \text{diag}[k_{11}, k_{21}, k_{31}], \tag{6.113}$$

$$\mathbf{K}_{j2} = \text{diag}[k_{12}, k_{22}, k_{32}] \tag{6.114}$$

and $\mathbf{J}'_4$ is the Jacobian matrix of the passive constraining leg with virtual joints.

The comparison between the parallel mechanism with rigid links (without virtual joints) and the parallel mechanism with flexible links (with virtual joints) is given in Table 6.4. The Cartesian compliance in each of the directions is given for a reference configuration of the mechanism, for progressively increasing values of the link stiffnesses.

From Table 6.4, one can find that with the improvement of link stiffness, the mechanism's compliance with flexible links is very close to that of mechanism with rigid links. This means that one can assume the flexible mechanism to be rigid only if the link stiffness reaches a high value.

6.4 Conclusions

In this chapter, mechanisms with revolute actuators (whose degrees of freedom are 3, 4, 5 and 6) have been considered. Solutions for the inverse kinematic problem have been given. The Jacobian matrices obtained have been used to establish the kinetostatic model of the mechanisms. The lumped models of the link and joint compliances have been used for the study of the Cartesian compliance. It has been shown that the kinetostatic analysis can be used to assess the stiffness properties of this family of mechanisms. Finally, examples have been investigated and numerical results have been obtained and the results clearly demonstrate the relevance of the kinetostatic analysis in the context of design of such mechanisms.

Chapter 7
Reconfigurable Parallel Kinematic Machine Tools

7.1 Preamble

The evolution of manufacturing systems is triggered by the dynamic customer environment of its time. The main characteristics of today's customers' environment are mass customization and responsiveness to market demand, and thus the reconfigurable manufacturing system has been suggested for such environment. A reconfigurable manufacturing system (RMS) is one designed at the outset for rapid change in its structure, as well as its hardware and software components, in order to quickly adjust its production capacity and functionality within a part family in response to sudden market changes or intrinsic system change [87]. Ideal reconfigurable manufacturing systems possess six main characteristics: Modularity, Integrability, Customized flexibility, Scalability, Convertibility, and Diagnosability (US patent, No. 6,349,237). These characteristics provide a RMS with exactly the functionality and production capacity needed, and also the system can be economically adjusted exactly when needed [105].

The components of RMS are: CNC machines [86], Reconfigurable Machine Tools [90], Reconfigurable Inspection Machines (US patent No. 6,567,162), and material transport systems (such as gantries and conveyors) that connect the machines to form the system. As the main component of reconfigurable manufacturing systems, the reconfigurable machine tools are machine tools that are built from machine modules [46]. Therefore, research and development in reconfigurable robots can generally be divided into two categories. One studies the most suitable modular architecture for robots. This includes the development of independent joint modules with various specifications and link modules as well as rapid interfaces between joints and links. The other is aimed at providing a CAD system for rapid formulation of a suitable configuration through a combination of those modular joints and links – a modular robot in its best conformity to a given task. In this chapter, first, we give some general idea about design procedures of reconfigurable parallel robotic machine tools, and then focus on the design of reconfigurable machine tools.

A machine module could be an actuator, a joint, a link, a tool holder, or a spindle. These modules are designed to be easily reconfigured to accommodate

D. Zhang, *Parallel Robotic Machine Tools*, DOI 10.1007/978-1-4419-1117-9_7,

new machining requirements. Therefore, the reconfigurable machine tools have the characteristics of modularity, convertibility, integratebility, customization, flexibility, and cost effectiveness.

The main objective of reconfigurable machine tool design is reconfigurable components that are reliable and costeffective. It is noted that machine design will be conducted at the component level. Some researchers already focused on the RMT study. Koren and Kota (US patent 5,943,750) received a patent for an RMT where the type, location, and number of spindles could be adjusted in response to changing product requirements. Landers et al. [90, 91] presented an overview of reconfigurable machine tools and their characteristics. Zhang et al. [170, 172, 173] developed a generic kinetostatic model for RMT stiffness analysis and design optimization. Yigit and Ulsoy [164] discussed the vibration isolation of RMTs and proposed different isolation strategies based on the RMT requirement.

As described by Koren et al. [87], the modular construction of parallel robotic machine, allows it to be used as a special class of reconfigurable machine tools, consists of simple and identical modules that can be configured into different machines. Parallel robotic machines can be also used as particular modules for machining lines.

According to Jovane [82], reconfigurable parallel robotic machines should have:

- Modular and reconfigurable structure with the possibility to change the total degree of freedom (DOF)
- Light structure and high dynamic performances, as a standard parallel robotic machine
- Possibility to use innovative materials in the machine design
- Possibility to use fast actuators such as linear motors
- Due to the intrinsic modularity, a DOF decoupling is recommended
- Low complexity in the kinematic chains

7.2 Theoretical Design

In order to maintain production process to be optimal, the machining systems must be adapted to the new demands. The adaptation process is called reconfiguration. Reconfiguration is the modification of the structure, the capacity, and technology at a later date. This is achieved by replacing, complementing, or removing self-contained functional modules. A reconfigurable machine tool (RMT) design has three phases (1) requirement definitions, (2) configuration generation, and (3) configuration selection. In addition, one should develop the machine modules for an RMT before the design process starts. The idea of RMT goes beyond the concept of modularity in which a RMT allows mass customization, facilitates easy integration of new technologies, is cost-effective, and provides high-speed capability. Systematic design tools have been developed for RMT. Zatarain et al. [168] propose a method to analyze the dynamic stiffness of a machine tool using precalculated component information. Roberto Pérez R. et al. [116] applies a concurrent design

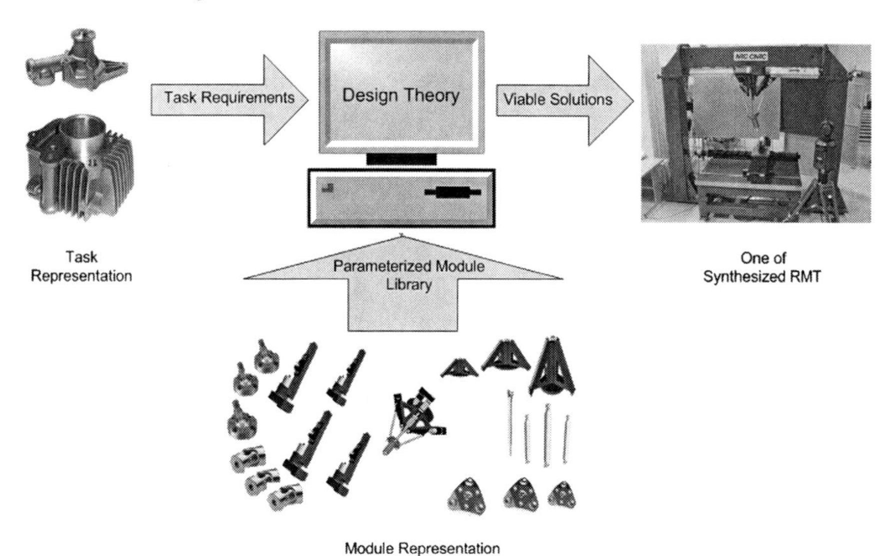

Fig. 7.1 Overview of RMT design methodology

reference model to RMT development. Wei Liu et al. present a methodology to optimize RMT design and warehouse based on a modified fuzzy-Chebyshev programming approach. Moon and Kota [109] have introduced the design methodology for RMT development. With the method, an example of such RMT that has been developed by the author and his colleagues at National Research Council of Canada is shown in Fig 7.1.

The machining operation is transformed into a task matrix (i.e., a homogeneous transformation matrix (HTM)) that contains the necessary motion requirements for the machine tool. The functional requirements of the machining operation are used to generate graph representations of candidate machine tools. A graph gives the overall topology of the machine tool and structural and kinematic functions are assigned to various portions of the graph. A library of machine tool modules (e.g., spindles, slides) containing structural and kinematic information for each module via HTMs, as well as connectivity information, is examined. Modules are assigned to various portions of the graph. The product of their HTMs is compared to the task matrix. If these matrices are equal, then the machine tool is kinematically viable. In this manner, all possible configurations can be determined. The viable configurations will be further reduced by other criteria (e.g., DOF, static and dynamic stiffness, thermal growth characteristics). This methodology also determines which modules must be added or deleted for each part in the part family.

In short, the key feature of this methodology is the use of screw-theory-based mathematical representation to transform a given description of machining tasks to be performed (process planning data) into a machine tool concept that is capable of performing the prescribed tasks. Starting from machining operations data, a set of feasible structural configurations of the machine is determined using graph theory. Various kinematic functions, (motions and base position) are then mapped to individual entities in each structural configuration. Using a precompiled parameterized library of commercially available machine modules, each function is then mapped to a feasible set of modules. This provides a set of kinematically feasible machine tools that provide desired motions.

7.3 Kinematics Model

In order to design the reconfigurable parallel kinematic machines properly, there are several criteria that need to be followed with, here are some major issues:

- The degree of freedom of the architecture
- Design constraints used for most promising typology selection
- Kinetostatic model for stiffness analysis and design optimisation

Mobility. A preliminary evaluation of the mobility of a kinematic chain can be found from the Chebychev–Grübler–Kutzbach formula.

$$M = d(n - g - 1) + \sum_{i=1}^{g} f_i, \quad (7.1)$$

Where M denotes the mobility or the system DOF, d is the order of the system ($d = 3$ for planar motion and $d = 6$ for spatial motion), n is the number of the links including the frames, g is the number of joints, and f_i is the number of DOFs for the ith joint.

Constraints. In order to ensure the required motions (i.e., 5-dof between the machine tool and the workpiece), the DOFs distribution numbers and the type of motions for each leg should be properly selected. Each leg can be facilitated with spherical, prismatic, Hooke and revolute joints.

The legs used in machine tools must be simple and practical to implement. For the sake of the simplicity and dexterity of mechanism, we prefer to use "spherical" pairs as the joints between link and the moving platform for those legs with more than 3 dofs. Since the serially connected revolute joints easily lead to singularities and the manufacturability is difficult, so we forgo their use for cases where more than 2 revolute joints connected in series.

This book focus on fully parallel mechanisms, but one can add legs (with 6-dof) to keep the structure symmetric. For the shape of the platforms, one should avoid the use of regular polygon, since it may lead to architecture singularities.

There are four basic types of joints, they are: revolute joint, Hook (universal) joint, Spherical joint, and prismatic joint (Fig. 7.2). One can develop the reconfigurable parallel robotic machine by combination of those joints and links.

In Figs. 7.3–7.5, we proposed a series of reconfigurable parallel mechanisms which consist of n identical actuated legs with six degrees of freedom and one

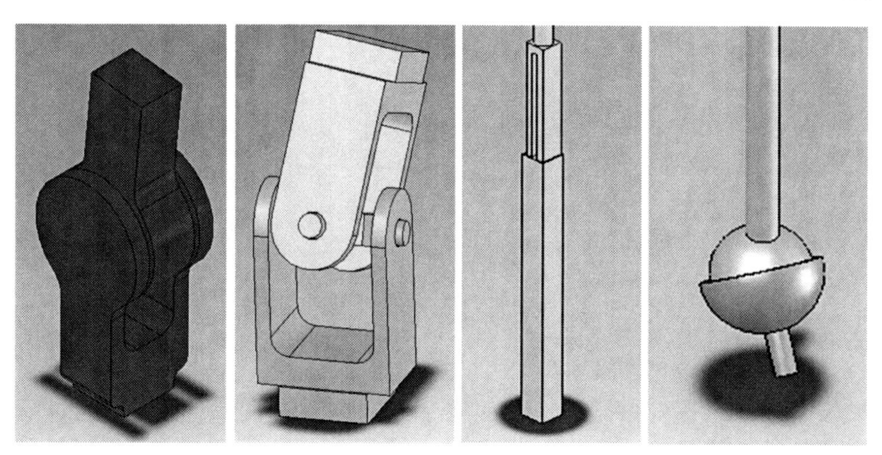

Fig. 7.2 Basic modular components

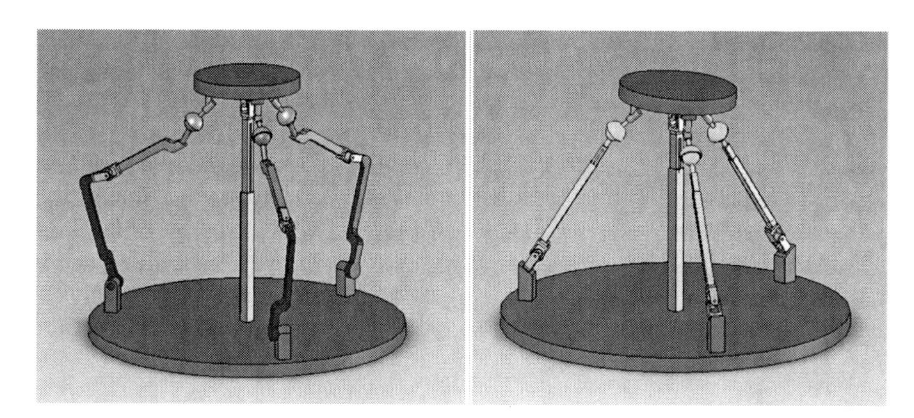

Fig. 7.3 Reconfigurable parallel robot with 3dof

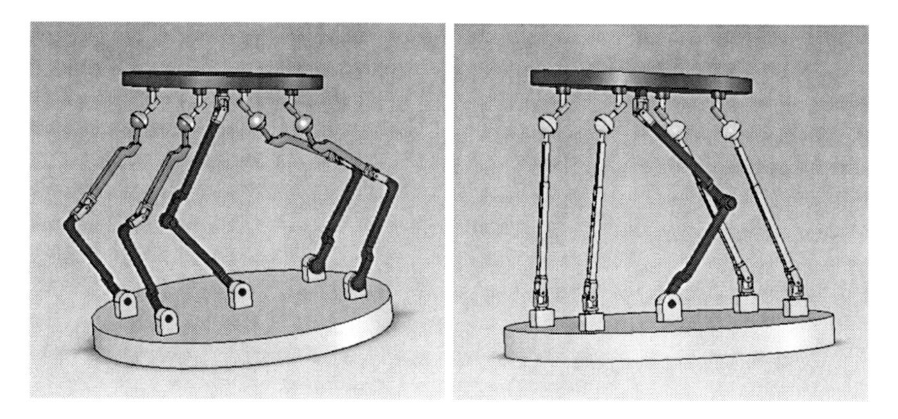

Fig. 7.4 Reconfigurable parallel robot with 4dof

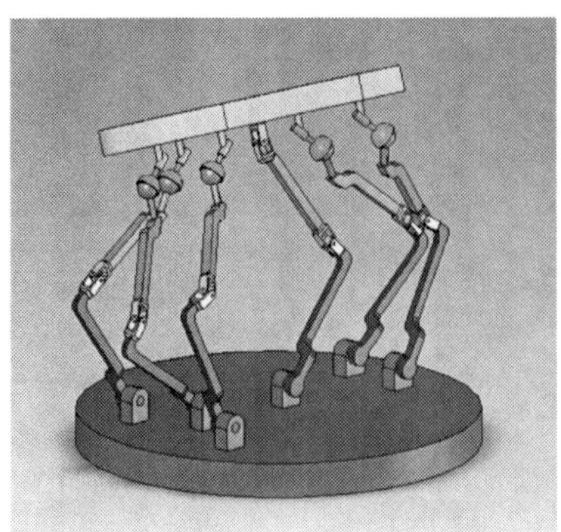
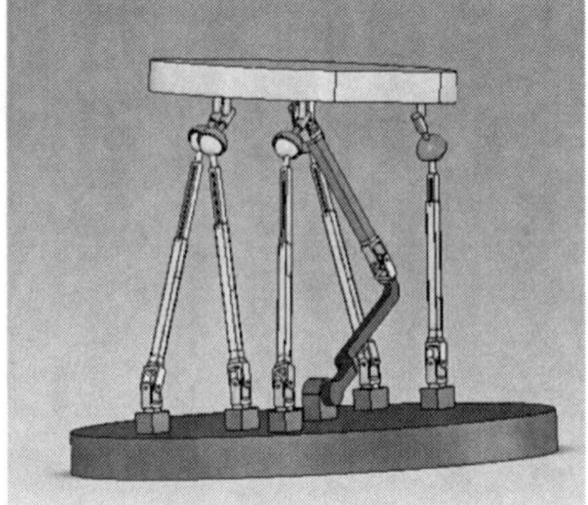

Fig. 7.5 Reconfigurable parallel robot with 5dof

passive leg with n degrees of freedom connecting the platform and the base. The degree of freedom of the mechanism is dependent on the passive leg's degree of freedom. One can improve the rigidity of this type of mechanism through optimization of the link rigidities, so as to reach a maximized global stiffness and precision. Finally, this series of mechanisms have the characteristics of reproduction since they have identical actuated legs, thus, the entire mechanism essentially consists of repeated parts, offering price benefits for manufacturing, assembling, and purchasing. It acts as a kind of reconfigurable parallel machines, based on a number of identical active kinematic chains plus a passive central leg, to compensate torsion. The reconfiguration process consists on the possibility to change the total DOF by adding or removing active struts, and, consequently constraining the passive leg.

7.4 Case Study

Figure 7.6 shows a 5-dof reconfigurable parallel robotic machine. Reconfigurable architectures are implemented by adjusting the three guideways and the combination with the X- and Y-table modules. This particular design employs an X–Y table to produce two degrees of freedom. These are translations along both the X- and Y-axes. The actual mechanism accounts for the other three degrees of freedom, i.e., translation along the Z-axis and rotations about the X- and Y-axes. These motions are produced by three motors which drive each of the three sliding rods. Each rod has a slider at both ends which moves along the slots in the base structure in order to guide the sliding rods. Additionally, connected to each sliding rod is a revolute joint which may also freely move along the rod. This motion is due to the fact that a slider which is attached to the revolute joint may move freely along a slot located in the sliding rod. Connected to each revolute joint are a moving link and a spherical joint. Each of these three "legs" is attached to the moving platform. The moving platform actually represents a main spindle which may hold different ma-

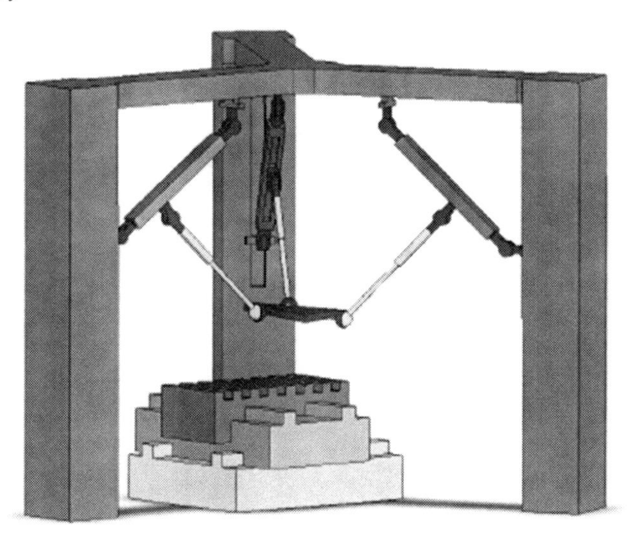

Fig. 7.6 Generic configuration of the 5-axis reconfigurable parallel kinematic machine

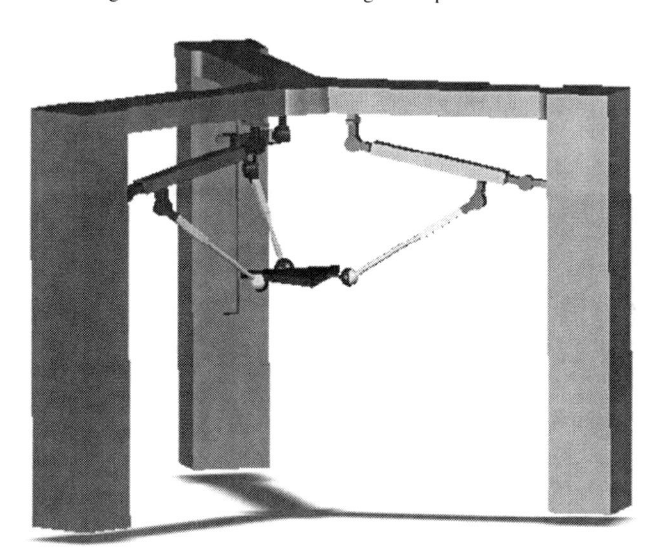

Fig. 7.7 Horizontal configuration of the 5-axis reconfigurable parallel kinematic machine

chining tools, such as a drill, milling cutter, lathe tool, etc. Thus, a work piece is placed upon the X–Y table and the tool is able to machine all sides of the work piece. There are four variations of this machine. The first (Fig. 7.6) is the generic configuration where each of the sliding rods is approximately at a 45° angle to the horizontal and vertical slots. The second configuration (Fig. 7.7) is the horizontal position. Here, the sliding rods are essentially in a horizontal position and therefore, the system is at a maximum stiffness. The third (Fig. 7.8) is the vertical position

Fig. 7.8 Vertical configuration of the 5-axis reconfigurable parallel kinematic machine

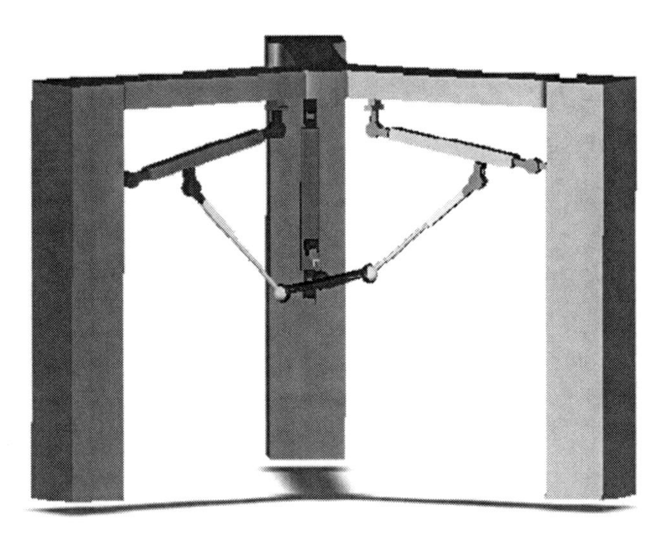

Fig. 7.9 Combination configuration of the 5-axis reconfigurable parallel kinematic machine

where the sliding rods are in a vertical position. This allows the system maximum access to the sides of a work piece; however, the system is at a minimum stiffness. The fourth configuration (Fig. 7.9) is a combination position. In this arrangement, two of the sliding rods are horizontal while one is vertical. This position is also used when work is required on the sides of a work piece.

7.5 Conclusions

The design of reconfigurable parallel robotic machine tools has been addressed in the chapter. Design methodology for this type of machine tool, especially for reconfigurable parallel robotic machine tool is discussed in detail. Examples are given to show the procedures of design and implementation. Since reconfigurable parallel robotic machine is a new type of machine tool, there are many issues that need to be addressed before it can be adopted widely. These issues include technology, accuracy, logistics, and implementation issues. However, reconfigurable parallel robotic machine has great promising to allow manufacturers to upgrade manufacturing facilities at minimal cost and develop manufacturing systems and product design simultaneously.

Chapter 8
Performance Evaluation of Parallel Robotic Machines

8.1 Preamble

Global stiffness and optimal calibration are the two crucial issues for parallel robotic machines for their performance, since global stiffness is directly related to the rigidity and accuracy of a parallel robotic machine, while optimal calibration can effectively improve the performance of the parallel robotic machine. In this chapter, both issues will be introduced and discussed. An example of a novel 3DOF parallel robotic machine will be illustrated in the chapter to show the detail of how to implement the global stiffness evaluation and optimal calibration. The method introduced in this chapter is very generic and can be applied in all types of robotic systems.

8.2 Global Stiffness Evaluation

8.2.1 Case Study: A Novel Three Degrees of Freedom Parallel Manipulator

Unlike the most existing 3-DOF parallel manipulators, the 3DOF parallel manipulator shown in this chapter contains a hybrid and uncoupled motion. The objective of the new design is to improve the system stiffness, eliminate coupled motions at the reference point, and simplify the kinematic model and control.

As shown in Fig. 8.1, the new parallel manipulator includes two innovative features. First, the universal joint of the passive link is located on the moving platform rather than on the base platform, thus eliminating the motions along the x and y translations and the z rotation. Second, the reference point on the moving platform has hybrid and independent motions with x and y rotations and a z translation. The new manipulator has three platforms: the base platform, labeled as $B_1 B_2 B_3$, the middle platform, designated as $M_1 M_2 M_3$, and the moving platform, identified as $E_1 E_2 E_3$. The base platform is fixed on the ground, and the middle platform is used to support the path, $B_i M_i$, of the actuated links, $D_i E_i$. The moving platform is used to mount a tool and helps to support the passive link, which is joined to the middle

D. Zhang, *Parallel Robotic Machine Tools*, DOI 10.1007/978-1-4419-1117-9_8,

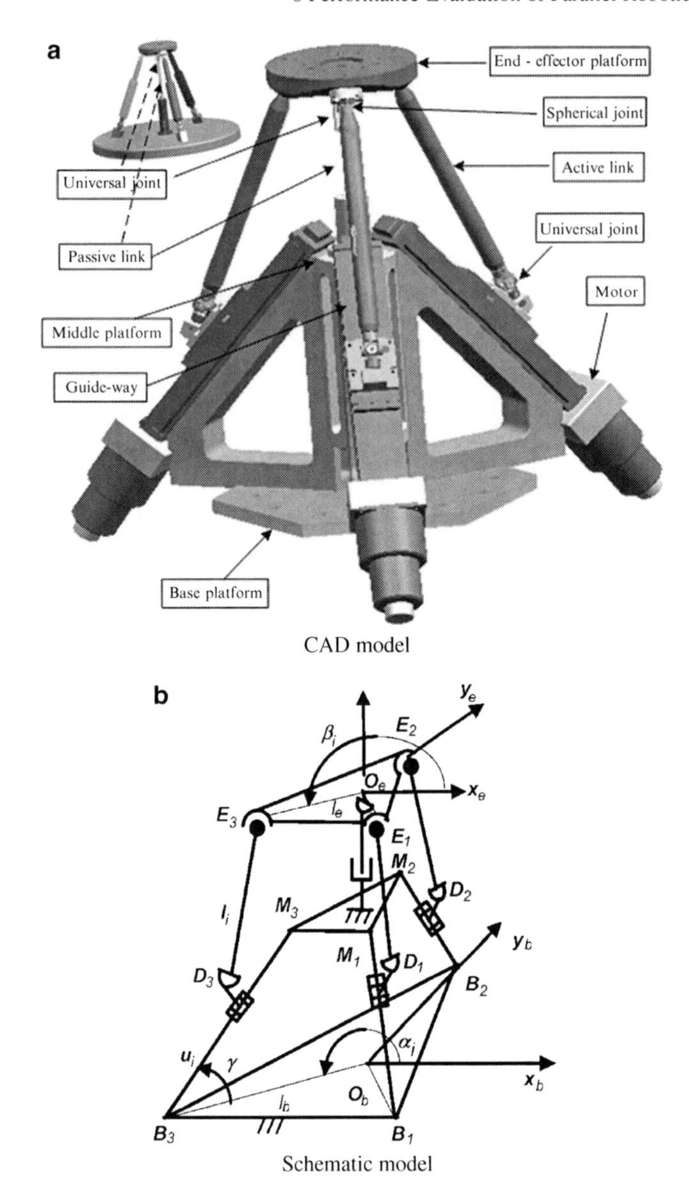

Fig. 8.1 The proposed 3-DOF parallel manipulator: (**a**) CAD model; (**b**) Schematic model

platform at the other end. The actuated links $D_i E_i$ are connected to the moving platform by a Spherical joint at E_i, and to a slider joined at the active ball screw by a universal joint at D_i. The passive link with the prismatic joint is fixed to the middle platform at one end and connected to the end-effector platform by a universal joint at the other end.

The following parameters define other details of the structure:

- The angle α_i (i=1,2,3) between x_b and $O_b B_i$
- The angle β_i (i=1,2,3) between x_e and $O_e E_i$
- The size of the base platform l_b
- The size of the end-effector platform l_e
- The direction of a guide-way γ
- The length of an active link l_i
- The offset of the spherical joints on the platform z_0

8.2.2 Kinematic Modeling

The 3-DOF manipulator is structured by two coordinate systems, $\mathbf{O}_e - \mathbf{x}_e\mathbf{y}_e\mathbf{z}_e$ and $\mathbf{O}_b - \mathbf{x}_b\mathbf{y}_b\mathbf{z}_b$, which are attached to the end-effector and base platforms, respectively.

For the origin of the end-effector, $\mathbf{O}_e$, the translational motions along $\mathbf{x}_e$ and $\mathbf{y}_e$ as well as the rotational motion along $\mathbf{z}_e$ are eliminated because of the use of the passive leg:

$$x_e = y_e = 0, \tag{8.1}$$

$$\theta_y = 0. \tag{8.2}$$

Therefore, the motions of $\mathbf{O}_e$ can be denoted by θ_x, θ_y, z_e, where θ_x and θ_y are the rotational motions along $\mathbf{x}_e$ and $\mathbf{y}_e$, and $\mathbf{z}_e$ is the translational motion along $\mathbf{z}_b$. The pose of the end-effector with respect to the coordinate system $\mathbf{O}_b - \mathbf{x}_b\mathbf{y}_b\mathbf{z}_b$ can be represented as

$$\mathbf{T}_e^b = \begin{bmatrix} \boldsymbol{R}_e & \boldsymbol{P}_e \\ 0 & \mathbf{1} \end{bmatrix} = \begin{bmatrix} c\theta_y & 0 & s\theta_y & 0 \\ s\theta_x s\theta_y & c\theta_x & -s\theta_x c\theta_y & 0 \\ -c\theta_x s\theta_y & s\theta_x & c\theta_x c\theta_y & z_e \\ 0 & 0 & 0 & 1 \end{bmatrix}, \tag{8.3}$$

where c and s denote the cosine and sine functions, respectively, $\mathbf{T}_e^b$ is the pose of the end-effector with respect to the coordinate system $\mathbf{O}_b - \mathbf{x}_b\mathbf{y}_b\mathbf{z}_b$, $\boldsymbol{R}_e$ is the 3×3 orientation matrix of the end-effector, and $\boldsymbol{P}_e$ is the location of $\boldsymbol{O}_e$.

The inverse kinematics can be formulated by finding the joint motions when the pose of the end-effector $\mathbf{T}_e^b$ is known. The joint motions are denoted by u_i, and the pose of the end-effector, $\mathbf{T}_e^b$, is determined by the motions of $\mathbf{O}_e(\theta_x, \theta_y, z_e)$. When solving the inverse kinematic problem, one can assume that the length of the support bar is constant.

The location of the connection between the end-effector platform and an active link is

$$\mathbf{P}_{e_i}^b = \boldsymbol{R}_e \mathbf{P}_{e_i}^e + \mathbf{P}_e^b = \begin{bmatrix} l_e c\beta_i c\theta_y + z_0 s\theta_y \\ l_e c\beta_i s\theta_x s\theta_y + l_e s\beta_i c\theta_x - z_0 s\theta_x c\theta_y \\ -l_e c\beta_i c\theta_x s\theta_y + l_e s\beta_i s\theta_x + z_e + z_0 c\theta_x c\theta_y \end{bmatrix}, \tag{8.4}$$

where

$$\mathbf{P}_{\mathrm{e}_i}^{\mathrm{b}} = \begin{bmatrix} x_{\mathrm{e}_i}^{\mathrm{b}} & y_{\mathrm{e}_i}^{\mathrm{b}} & z_{\mathrm{e}_i}^{\mathrm{b}} \end{bmatrix}^{\mathrm{T}},$$
$$\mathbf{P}_{\mathrm{e}_i}^{\mathrm{e}} = \begin{bmatrix} l_{\mathrm{e}} c\beta_i & l_{\mathrm{e}} s\beta_i & z_0 \end{bmatrix}^{\mathrm{T}}.$$

In (8.4), z_0 is the offset of the spherical joint with respect to $\boldsymbol{O}_{\mathrm{e}}$. Equation (8.4) can be differentiated as follows:

$$\begin{bmatrix} \delta x_{\mathrm{e}_i}^{\mathrm{b}} \\ \delta y_{\mathrm{e}_i}^{\mathrm{b}} \\ \delta z_{\mathrm{e}_i}^{b} \end{bmatrix} = [\mathbf{J}_i]_{3\times 3} \begin{bmatrix} \delta\theta_x \\ \delta\theta_y \\ \delta z_{\mathrm{e}} \end{bmatrix}, \tag{8.5}$$

where

$$\mathbf{J}_i = \begin{bmatrix} 0 & -l_{\mathrm{e}} c\beta_i s\theta_y + z_0 c\theta_y & 0 \\ (l_{\mathrm{e}} c\beta_i s\theta_y - z_0 c\theta_y) c\theta_x - l_{\mathrm{e}} s\beta_i s\theta_x & (l_{\mathrm{e}} c\beta_i c\theta_y + z_0 s\theta_y) s\theta_x & 0 \\ (l_{\mathrm{e}} c\beta_i s\theta_y - z_0 c\theta_y) s\theta_x + l_{\mathrm{e}} s\beta_i c\theta_x & -(l_{\mathrm{e}} c\beta_i c\theta_y + z_0 s\theta_y) c\theta_x & 1 \end{bmatrix}.$$

Since the active links have a fixed length, it can be shown that

$$|\, \mathbf{O}_{\mathrm{b}}\mathbf{E}_i - \mathbf{O}_{\mathrm{b}}\mathbf{B}_i - \mathbf{B}_i\mathbf{D}_i \,|=|\, \mathbf{D}_i\mathbf{E}_i \,|\ (i = 1, 2, 3) \tag{8.6}$$

Equation (8.6) yields

$$k_{i1}^2 + k_{i2}^2 + k_{i3}^2 = l_i^2, \tag{8.7}$$

where

$$k_{i1} = x_{\mathrm{e}_i}^{\mathrm{b}} - (l_{\mathrm{b}} - u_i c\gamma) c\alpha_i,$$
$$k_{i2} = y_{\mathrm{e}_i}^{\mathrm{b}} - (l_{\mathrm{b}} - u_i c\gamma) s\alpha_i,$$
$$k_{i3} = z_{\mathrm{e}_i}^{\mathrm{b}} - u_i s\gamma.$$

Assuming that there is only linear motion in the actuator of each active link, that the active link is a two-force component, and that only axial deformation occurs, then (8.7) can be differentiated as:

$$\begin{bmatrix} \delta \mathbf{u}_i \\ \delta \mathbf{l}_i \end{bmatrix} = \begin{bmatrix} -\dfrac{k_{i1}}{k_{i4}} & -\dfrac{k_{i2}}{k_{i4}} & -\dfrac{k_{i3}}{k_{i4}} \\ \dfrac{k_{i1}}{l_i} & \dfrac{k_{i2}}{l_i} & \dfrac{k_{i3}}{l_i} \end{bmatrix} \cdot \begin{bmatrix} \delta x_{\mathrm{e}_i}^{\mathrm{b}} \\ \delta y_{\mathrm{e}_i}^{\mathrm{b}} \\ \delta z_{\mathrm{e}_i}^{\mathrm{b}} \end{bmatrix} (i = 1, 2, 3), \tag{8.8}$$

where

$$k_{i4} = k_{i1} c\gamma c\alpha_i + k_{i2} c\gamma s\alpha_i - k_{i3} s\gamma.$$

Substituting (8.5) into (8.8),

$$\begin{bmatrix} \delta u_1 \\ \delta u_2 \\ \delta u_3 \end{bmatrix} = [\mathbf{J}_t]_{3\times 3} \begin{bmatrix} \delta\theta_x \\ \delta\theta_y \\ \delta z_e \end{bmatrix} = \begin{bmatrix} (J_{t,1})_{1\times 3} \\ (J_{t,2})_{1\times 3} \\ (J_{t,3})_{1\times 3} \end{bmatrix} \begin{bmatrix} \delta\theta_x \\ \delta\theta_y \\ \delta z_e \end{bmatrix}, \tag{8.9}$$

where

$$\mathbf{J}_{t,i} = \begin{bmatrix} -\frac{k_{i1}}{k_{i4}} & -\frac{k_{i2}}{k_{i4}} & -\frac{k_{i3}}{k_{i4}} \end{bmatrix} \cdot \mathbf{J}_i .$$

The twist of the platform can be defined as

$$\mathbf{t} = \begin{bmatrix} \boldsymbol{\omega}^{\mathrm{T}} & \dot{\mathbf{p}}^{\mathrm{T}} \end{bmatrix}^{\mathrm{T}} = \begin{bmatrix} \delta\theta_x \; \delta\theta_y \; \delta z_e \end{bmatrix}, \tag{8.10}$$

i.e.,

$$\dot{\rho} = \mathbf{J}\mathbf{t} = \begin{bmatrix} \delta l_1 & \delta l_2 & \delta l_3 \end{bmatrix}^{\mathrm{T}}$$

$$\begin{bmatrix} \delta l_1 \\ \delta l_2 \\ \delta l_3 \end{bmatrix} = [\mathbf{J}_a]_{3\times 3} \begin{bmatrix} \delta\theta_x \\ \delta\theta_y \\ \delta z_e \end{bmatrix} = \begin{bmatrix} (J_{a,1})_{1\times 3} \\ (J_{a,2})_{1\times 3} \\ (J_{a,3})_{1\times 3} \end{bmatrix} \begin{bmatrix} \delta\theta_x \\ \delta\theta_y \\ \delta z_e \end{bmatrix},$$

where

$$\mathbf{J}_{a,i} = \begin{bmatrix} \frac{k_{i1}}{l_i} & \frac{k_{i2}}{l_i} & \frac{k_{i3}}{l_i} \end{bmatrix} \cdot \mathbf{J}_i .$$

8.2.3 *The Global Stiffness Evaluation*

In this section, a practical case will be examined for the parameters of the manipulator provided in Tables 8.1 and 8.2.

Using the kinematic model, the workspace is simulated, as shown in Fig. 8.2. The working ranges of θ_x, θ_y, and z_e are $(-40°, 40°)$, $(-40°, 40°)$, and $(0.440, 0.608)$, respectively.

Table 8.1 Kinematic parameters

α_i	$(-30°, 90°, -150°)$
β	$(-30°, 90°, -150°)$
l_b	374.25 mm
l_e	75 mm
l_i	353 mm
γ	50°
Z_{e0}	13 mm
Prismatic joint (active)	240.272 ± 85 mm
Spherical joint (passive)	$(-45°, 45°)$
Universal joint (passive)	$(-50°, 50°)$

Table 8.2 Dynamic parameters

	Mass (kg)	Mass center	Moment of inertia (kg m^2) (I_{xx}, I_{yy}, I_{zz})
Moving platform			
Total	1.0	$\mathbf{O}_e$	(0.0, 0.0, 0.0)
Platform	0.38	$\mathbf{O}_e$	(0.002830, 0.002123, 0.004953)
Support bar	0.1945	$\mathbf{F}_i$	(0.000717, 0.000717, 0.000010)
Slider	0.3045	$\mathbf{D}_i$	
Passive leg	0.2	$\mathbf{P}$	

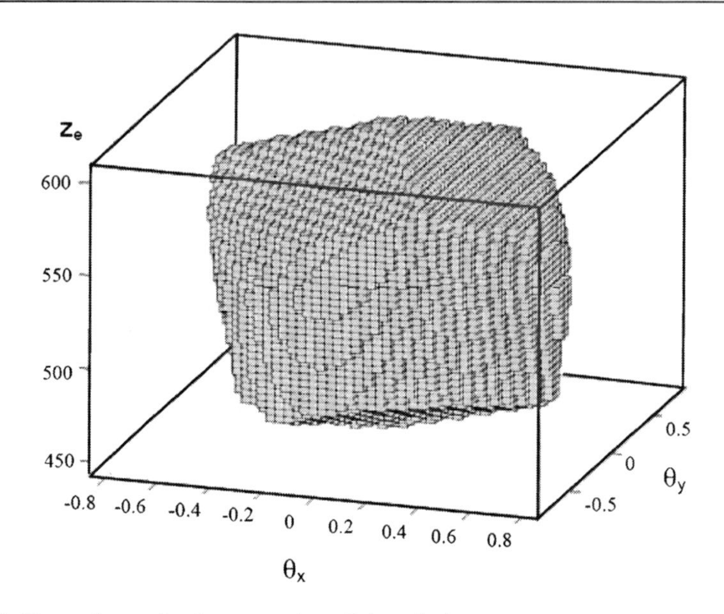

Fig. 8.2 The workspace for the proposed parallel manipulator

The calculation methodology of global stiffness is similar with the general cases studied the previous Chaps. 5 and 6. In performing the system stiffness simulation, the stiffness coefficients of the links are calculated as:

$$
\begin{aligned}
k_t &= 1.5177 \times 10^{10}, \\
k_a &= 9.5950 \times 10^{7}, \\
k_{\theta x} &= 899.6698/(z_{\mathrm{e}} - z_{\mathrm{b}})^3, \\
k_{\theta y} &= 899.6698/(z_{\mathrm{e}} - z_{\mathrm{b}})^3, \\
k_{\theta z} &= 697.4185/(z_{\mathrm{e}} - z_{\mathrm{b}})^2.
\end{aligned}
$$

Note that stiffness coefficients depend on the z-coordinate of the end-effector platform.

Over this cross-section, the results of the evaluation are shown in Table 8.3. The stiffness of the motion and constrained axes are illustrated in Fig. 8.3. Stiffness is an

Table 8.3 Stiffness over the section $z_e = (min\_z_e + max\_z_e)/2$

Principal axis	Minimum	Maximum	Average
Actuated links			
θ_x	5.6056×10^7	1.3026×10^8	1.1461×10^8
θ_y	4.5293×10^7	1.3140×10^8	1.1088×10^8
T_z	3.6701×10^{10}	3.7756×10^{10}	3.7464×10^{10}
Passive link			
T_x	6.0793×10^7		
T_y	6.0793×10^7		
θ_z	5.9621×10^4		
Global	6.1633×10^9	6.3544×10^9	6.3015×10^9

important factor in the design of machine tools, since it affects the precision of the machining. Specifically, global stiffness, illustrated in Fig. 8.4, is directly related to system rigidity.

From the preceding analysis of mechanical stiffness, a number of conclusions can be drawn. First, the distribution of stiffness varies between the motion axes and the constrained axes. The stiffness on the motion axes is the greatest at the center position $\theta_x = \theta_y = 0$, and the stiffness on the constrained axes is the same over the cross-section in the case of the z-coordinate. Second, the stiffness on the z translation is the greatest, and it is the weakest on the z rotation. Finally, the distribution of global stiffness is similar to the distribution of stiffness on a motion axis, where the center position possesses the highest value.

The stiffness for the weakest z rotation axis can be enhanced by increasing the radius of the passive link. Since the stiffness is proportional to the stiffness coefficient of the component, the stiffness on the constrained axes is completely determined by the passive link.

8.3 Optimal Calibration Method

Error is one of the crucial factors, which affects the performance indices of parallel robots. According to the sources and characteristics, the error of parallel robots can be classified as in Table 8.4. For the error sources of parallel robots, it can be divided into two main types-vibration error and quasistatic error. Vibration error is the operational inaccuracy induced by the vibration of tools, system axis, or motor. Quasistatic error includes the kinematic parameter error (or manufacturing and assembly error), thermal error, and nonlinear stiffness error.

This differs from many studies reported in the past literature. The thermal error induced in the sliding motion (i.e., actuator leg) is a major error source in the operation of the end-effector [4], and it is distributing over the entire structure of a robotic mechanism. The nonlinear stiffness error may be caused by materials of links and joints and/or external forces and/or moments which induce deformation

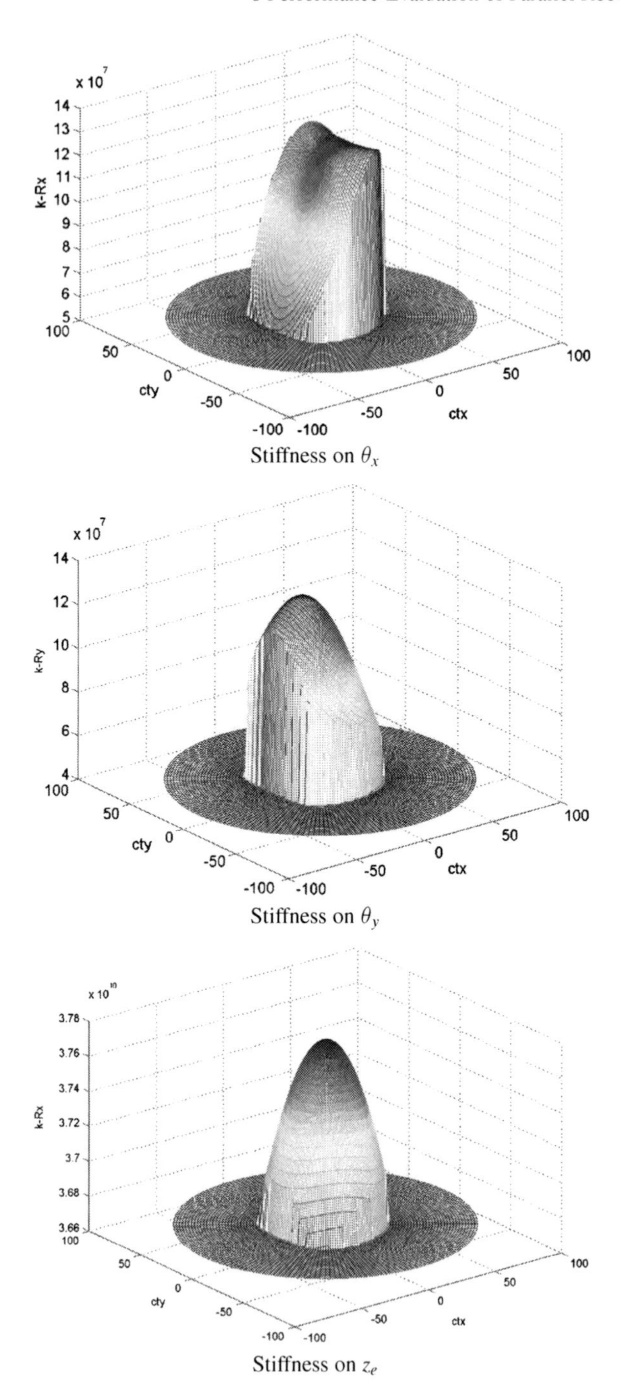

Fig. 8.3 Stiffness on the cross-section $z_e = (min\_z_e + max\_z_e)/2$: (**a**)Stiffness on θ_x; (**b**)Stiffness on θ_y; (**c**)Stiffness on z_e

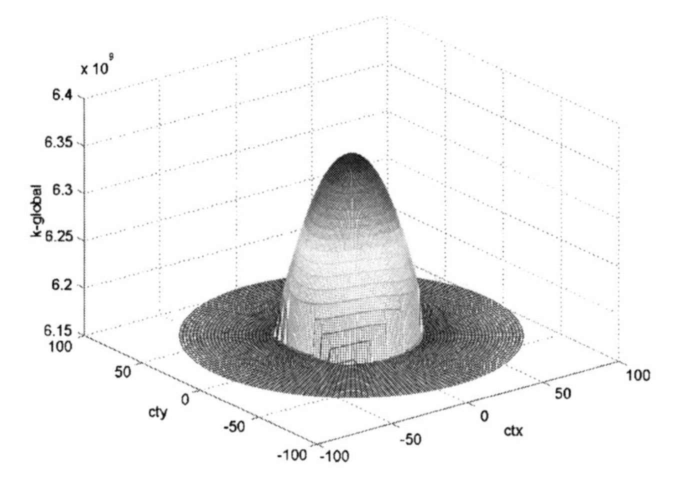

Fig. 8.4 Global stiffness over the cross-section $z_e = (min\_z_e + max\_z_e)/2$

Table 8.4 The error classification of parallel robots

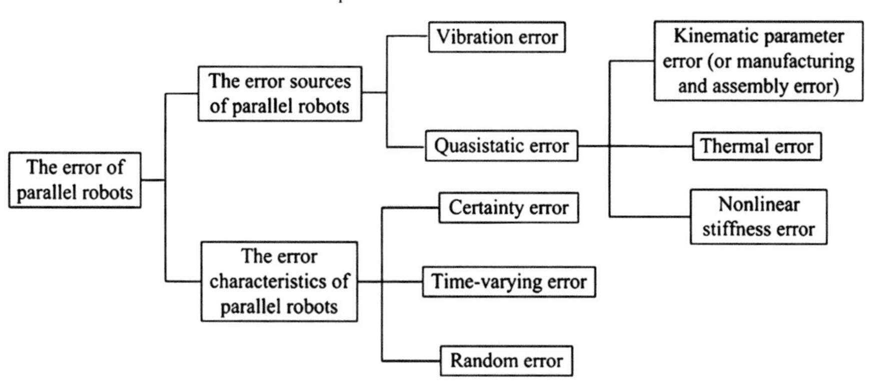

(large deformation or small deformation). The errors over an entire workspace can be visualized to guide effectively the design process. It can be expressed with the notion of an error map (analogous to the stiffness map, as mentioned earlier). The error map will allow an active participation of designers in determining strategies for error compensation and understanding sources of errors (i.e., manufacturing and assembly, thermal, or nonlinear stiffness) to determine a way to remove/reduce the errors.

In this section, we will discuss how to conduct effective error compensation and calibration, thereby reducing its impact to the operational accuracy. The source of manipulator error is in a wide range. In addition to the main factors, such as the error induced by geometric size and thermal effect, there are many other factors, e.g., load deformation, gear clearance, voltage fluctuations external environment, etc. So it is very difficult to establish a general error model considering all the error sources. On

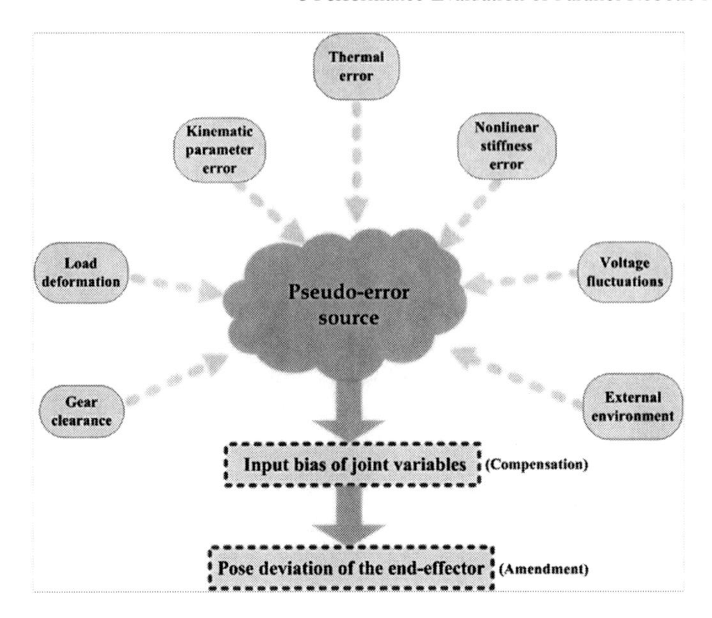

Fig. 8.5 Relationship of the pseudo-error source and the error of end-effector

the basis of the method of pose compensation of the end-effector, the multi-error sources included manufacturing and assembly error, and thermal effect error and nonlinear stiffness error can be abstracted as a single error source, namely pseudo error source. The relationship of the pseudo-error source and the end-effector error is shown in Fig. 8.5.

Assuming that the ideal pose the end-effector of parallel robot with respect to the fixed reference frame can be defined as

$$\mathbf{Pose}_{\text{real}} = \{x, y, z, \psi, \theta, \phi\}, \tag{8.11}$$

where, the first three elements express position values, and the last three elements express orientation values. The homogeneous matrix of the end-effector with respect to the fixed reference frame can be expressed by $\mathbf{T}_{\text{P}}^{\text{O}}$,

$$\mathbf{T}_{\text{P}}^{\text{O}} = \begin{bmatrix} c\psi c\theta & c\psi s\theta s\phi - s\psi c\phi & c\psi s\theta c\phi + s\psi s\phi & x \\ s\psi c\theta & s\psi s\theta s\phi + c\psi c\phi & s\psi s\theta c\phi - c\psi s\phi & y \\ -s\psi & c\theta s\phi & c\theta c\phi & z \\ 0 & 0 & 0 & 1 \end{bmatrix}. \tag{8.12}$$

When the real pose is not equal to the ideal value, it has

$$\begin{aligned} \mathbf{Pose}_{\text{real}} &= \{x', y', z', \psi', \theta', \phi'\} \\ &= \{x + \Delta x, y + \Delta y, z + \Delta z, \psi + \Delta\psi, \theta + \Delta\theta, \phi + \Delta\phi\}, \end{aligned} \tag{8.13}$$

where, $\mathbf{Pose}_{\text{real}}$ is the real pose of the end-effector with respect to the fixed reference frame. Thus, when error happens, the homogeneous matrix $\mathbf{T}^{\text{O}}_{\text{P(Error)}}$ can be obtained as

$$\mathbf{T}^{\text{O}}_{\text{P(Error)}} = \begin{bmatrix} c\psi' c\theta' & c\psi' s\theta' s\phi' - s\psi' c\phi' & c\psi' s\theta' c\phi' + s\psi' s\phi' & x' \\ s\psi' c\theta' & s\psi' s\theta' s\phi' + c\psi' c\phi' & s\psi' s\theta' c\phi' - c\psi' s\phi' & y' \\ -s\psi' & c\theta' s\phi' & c\theta' c\phi' & z' \\ 0 & 0 & 0 & 1 \end{bmatrix}. \quad (8.14)$$

The energy function of pose error of the end-effector can be expressed as

$$\mathbf{E} = \mathbf{E}_{xyz} + \eta \cdot \mathbf{E}_{\psi\theta\phi}, \quad (8.15)$$

where, position error is defined by Euclidean distance, namely

$$\mathbf{E}_{xyz} = (\Delta x^2 + \Delta y^2 + \Delta z^2)^{1/2}. \quad (8.16)$$

Likelihood, orientation error can be expressed by cosine:

$$\mathbf{E}_{\psi\theta\phi} = cos^{-1}(c\psi \cdot c(\psi + \Delta\psi) + c\theta \cdot c(\theta + \Delta\theta) + c\phi \cdot c(\phi + \Delta\phi)). \quad (8.17)$$

The system stiffness model and the neural network approach can be applied for the calibration process. It is known that calibration is best performed in the least sensitive error region within an entire workspace. Furthermore, calibration should be conducted with the minimum number of joint configurations, as the calibration

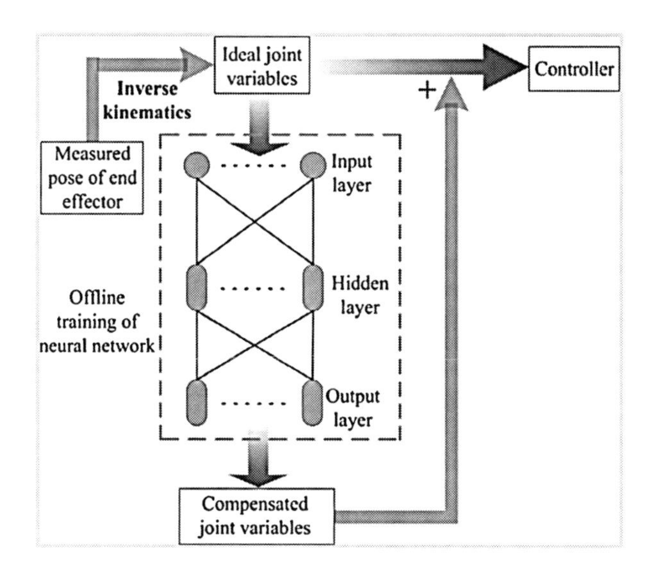

Fig. 8.6 The error compensation flow

process is extremely time-consuming. The system stiffness model is based on the kinetostatic model, with which the explicit expression of system stiffness can be generated at any pose. Thus, the critical components that have the largest effect on system stiffness can be identified. Because of the complexity of the error sources, it is difficult to develop the calibration model if all the errors will be considered. In this book, errors including manufacturing and assembly error, thermal error, and nonlinear stiffness error are considered as a single error source (pseudo-error source), which only causes the deviation of joint variables. The neural network can be applied to describe the complex nonlinear relationship between joint variables (control parameters) and deviation of joint variables with respect to the measured pose of the end-effector. With neural network, the pseudo-error in arbitrary joint variable can be obtained and thus the control parameters can be adjusted accordingly (as shown in Fig. 8.6). This approach is generic and feasible for all types of robotic system. The method has been validated by the result reported by Vener et al. [149].

8.4 Conclusions

This chapter demonstrated how to conduct the global stiffness evaluation and to improve the parallel robotic machine through optimal calibration. A unique 3-DOF parallel robotic machine with pure 3-DOF motion is taken as an example. Through this chapter, people can implement the performance evaluation and improvement effectively using the introduced method.

Chapter 9
Design Optimization of Parallel Robotic Machines

9.1 Preamble

Optimization plays an important role in engineering design problems; it deals with problems of minimizing or maximizing a function with several variables. The purpose of optimization design is aiming at enhancing the performance indices by adjusting the structure parameters such as link length, radii of fixed platform and moving platform, and its distance between the center points of the two platforms. The approach can been called dimensional-synthesis-based performance optimization of parallel manipulator. In the optimum design process, several performance criteria could be involved for a design purpose, such as stiffness, dexterity, accuracy, workspace, etc.

Many researchers have studied on the issue of optimal design of robot manipulators [19, 89, 166, 178]. Zhao et al. [182] exploited the least number method of variables to optimize the leg length of a spatial parallel manipulator for the purpose of obtaining a desired dexterous workspace. Stock and Miller [138] presented a method for multidimensional kinematic optimization of the linear Delta robot architecture's geometry. A utility objective function was formulated incorporating two performance indices, including manipulability and space utilization. Rout and Mittal [131] proposed the experimental approach for the optimization of the dimension of 2-dof R–R planar manipulator. Kucuk and Bingul [89] optimized the workspace of two spherical three links robot manipulators using the local and global performance indices. Mitchell et al. [108] presented kinematic optimization to confirm the smallest configuration that would satisfy the workspace requirements for a lightweight and compact surgical manipulator. Chablat and Angeles [31] investigated on optimum dimensioning of revolute-coupled planar manipulators based on the concept of distance of Jacobian matrix to a given isotropic matrix which was used as a reference model. Boeij et al. [23] proposed numerical integration and sequential quadratic programming method for optimization of a contactless electromagnetic planar 6-dof actuator with manipulator on top of the floating platform. Ceccarelli and Lanni [29] investigated the multiobjective optimization problem of a general 3R manipulator for prescribed workspace limits and numerically using an algebraic formulation. As the primary components of artificial intelligence approach, genetic algorithms and

D. Zhang, *Parallel Robotic Machine Tools*, DOI 10.1007/978-1-4419-1117-9_9,

artificial neural networks play the important roles in various fields of science and technology. In this chapter, the two methods are applied as the optimization criteria for the synthesis of stiffness and other criteria.

The traditional optimization methods can only handle a few geometric variables due to the lack of convergence of the optimization algorithm. However, genetic algorithms have applied the powerful and broadly applicable stochastic search methods and optimization techniques, and they can escape from local optima [69]. Therefore, genetic algorithms have been selected as the best candidate to address the convergence issue and are suitable for performance optimization of the proposed mechanism in the previous several chapters. Neural networks possess the capability of complex function approximation and generalization by simulating the basic functionality of the human nervous system in an attempt to partially capture some of its computational strengths. Since the solution of objective function must be solved before using genetic algorithms, neural networks will be conducted to represent the objective functions of performance indices. On the investigation of multiobjective optimization problem in design process, since it is impossible to maximize or minimize all objective function values if they are conflicting with each other, the trade-off process should be executed. This methodology will pave the way for providing not only the effective guidance, but also a new approach of dimensional synthesis for the optimal design of general parallel mechanisms.

For the mechanisms studied here, the highest global stiffness are desired so as to reach the high rigidity and high precision. This can be achieved either through maximizing the global stiffness or through minimizing the global compliances for a certain parallel mechanism by selecting mechanism's geometric parameters (link length, height, etc.) and behavior parameters (link stiffness). In this chapter, the optimization criteria are first established. An optimization process based on genetic algorithms is applied for the global stiffness of all the proposed spatial parallel/hybrid mechanisms for 6dof to 3dof, and the rationale for using this method together with the determination of parameters and objective function are addressed as well. The detailed analysis of the kinetostatics of the parallel mechanisms conducted in previous chapters will now be used to define and optimize their geometric sizes and properties. Furthermore, for the issue of multiobjective optimization, two cases are investigated where the integration methodology of genetic algorithms and artificial neural networks is implemented to search the optimal architecture and behavior parameters in terms of various optimization objectives including global stiffness, dexterity, and manipulability.

9.2 Optimization Objective and Criteria

In this book, the main consideration for the optimization criteria is to maximize global stiffness (or minimize the global compliances). The global stiffness/compliance used here is the diagonal entry of the Cartesian stiffness/compliance matrix. It represents the pure stiffness/compliance in each direction. Genetic algorithm methods are used to conduct the optimal design of the system in terms of a better system stiffness. The objective functions are established and maximized/minimized in order to find the suitable geometric parameters (coordinates of the attachment points, coordinates of the moving platform, link length, vertex distributions at base and moving platform, platform height, etc.) and behavior parameters (actuator stiffness, actuated link stiffness, kinetostatic model stiffness, etc.) of the mechanisms. Since the objective function is closely related to the topology and geometry of the structure, the general optimization methodology can be described as follows:

- Analyze the requirements including the stiffness, the mechanical interferences, the workspace properties, and the singularities
- Analyze the constraints including geometric size and properties
- Establish a reasonable initial guess of the geometry of the mechanism, then use a numerical optimization to further improve the kinematic properties and ensure the optimum characteristics are obtained. Finally, a program gives a potential solution to allow the verification of other important properties.

9.3 Basic Theory of Evolutionary Algorithms

Introduced in the 1970s by John Holland [69], genetic algorithms are part of the larger class of evolutionary algorithms that also include evolutionary programming [49], evolution strategies [126], and genetic programming [88]. The genetic algorithms (GAs) are powerful and broadly applicable stochastic search and optimization techniques based on the evolutionary principle of natural chromosomes [53]. Specifically, the evolution of chromosomes due to the operation of crossover and mutation and natural selection of chromosomes based on Darwin's survival-of-the-fittest principles are all artificially simulated to constitute a robust search and optimization procedure. The genetic algorithms are the computer simulation of such evolution where the user provides the environment (function) in which the population must evolve.

A comparison between conventional optimization methods and genetic algorithms is now given. The conventional methods are usually limited to convex regular functions while the genetic algorithm is robust, global, and generally more straightforward to apply to all kinds of functions including multimodal, discontinuous, and nondifferentiable functions. Goldberg [53] has summarized the differences as follows:

1. Genetic algorithms work with a coding of the solution set, not the solutions themselves.
2. Genetic algorithms search from a population of solutions, not a single solution.
3. Genetic algorithms use payoff information (fitness function), not derivatives or other auxiliary knowledge.
4. Genetic algorithms use probabilistic transition rules, not deterministic rules.

In recent years, the GAs have been applied to a broad range of real-world problems [25, 26, 42, 43, 51, 54, 107, 110, 157] such as ecosystem modeling, combinatorial and parametric optimization, reliability design, vehicle routing and scheduling, machine intelligence, robotic trajectory optimization, neural networks implementations, pattern recognition, analysis of complex systems, and financial prediction.

The basic procedure of genetic algorithms can be described as follows:

1. Create an initial population: The initial population of chromosomes is created randomly.
2. Evaluate all of the individuals (apply some function or formula to the individuals) The fitness is computed in this step. The goal of the fitness function is to numerically encode the performance of the chromosomes.
3. Selection: Select a new population from the old population based on the fitness of the individuals as given by the evaluation function. In this step, the chromosomes with the largest fitness rates are selected while the chromosomes with low fitness rates are removed from the population.
4. Genetic operations (mutation and crossover): If the parents are allowed to mate, a recombination operator is employed to exchange genes between the two parents to produce two children. If they are not allowed to mate, the parents are placed into the next generation unchanged. A mutation simply changes the value for a particular gene.
5. Evaluate these newly created individuals.
6. Repeat steps 3–5 (one generation) until the termination criteria have been satisfied.

Suppose $P(t)$ and $C(t)$ are parents and children in current generation t, then a genetic algorithm is expressed in Fig. 9.1. From Fig. 9.1, one can find that there are only two kinds of operations included in genetic algorithms, i.e., genetic operations (crossover and mutation) and evolution operation (selection).

```
begin
  t = 0;
  initialize P(t);
  evaluate P(t);
  while (unfinished condition) do
    select P ' (t) from P(t);
    reproduce C(t) from P ' (t);
    mutate C(t);
    evaluate  C(t);
    t = t + 1;
  end
end
```

Fig. 9.1 The structure of genetic algorithms

Genetic algorithms have the advantages of robustness and good convergence properties, namely:

- They require no knowledge or gradient information about the optimization problems. They can solve any kind of objective functions and any kind of constraints (i.e., linear or nonlinear) defined on discrete, continuous, or mixed search spaces.
- Discontinuities present on the optimization problems have little effect on the overall optimization performance.
- They are effective at performing global search (in probability) instead of local optima.
- They perform very well for large-scale optimization problems.
- They can be employed for a wide variety of optimization problems.

Genetic algorithms have been shown to solve linear and nonlinear problems by exploring all regions of state space and exponentially exploiting promising areas through mutation, crossover, and selection operations applied to individuals in the population [107].

In the present work, there are many optimization parameters (up to 13 variables, depending on mechanism, make up the optimization problem) and complex matrix computations. Hence, it is very difficult to write out the analytical expressions for each stiffness element. Moreover, with traditional optimization methods, only a few geometric parameters [60] could be handled due to the lack of convergence of the optimization algorithm when used with more complex problems. This arises from the fact that traditional optimization methods use a local search by a convergent stepwise procedure (e.g., gradient, Hessians, linearity, and continuity), which compares the values of the next points and moves to the relative optimal points. Therefore, genetic algorithms are the best candidate for the optimization problems studied here.

In order to use genetic algorithms properly, several parameter settings have to be determined, they are: chromosome representation, selection function, genetic operators, the creation of the population size, mutation rate, crossover rate, and the evaluation function.

1. Chromosome representation:
   This is a basic issue for the GA representation, it is used to describe each individual in the population of interest. In the original algorithm, each individual or chromosome used to be expressed as a sequence of genes from binary digits (0 and 1) [69]. However, it has been shown that more natural representations are more efficient and produce better solutions [107]. Michalewicz [107] has done extensive experimentation comparing real-valued and binary genetic algorithms and shows that the real-valued genetic algorithm is an order of magnitude more efficient in terms of CPU time. He also shows that a real-valued representation moves the problem closer to the problem representation which offers higher precision with more consistent results across replications [107]. It outperformed

binary genetic algorithm and simulated annealing in terms of computational efficiency and solution quality [124]. Hence, real-valued expressions are used in our case to represent each individual or chromosome for function optimization. For the problem studied here, the chromosomes consist of the architecture parameters (coordinates of the attachment points, coordinates of the moving platform, link lengths, vertex distributions at base and moving platform, platform height, etc.) and behavior parameters (actuator stiffness, actuated link stiffness, kinetostatic model stiffness, etc.) of the mechanisms.

2. Selection function:
   This step is a key procedure to produce the successive generations. It determines which of the individuals will survive and continue on to the next generation. A probabilistic selection is performed based on the individual's fitness such that the better individuals have an increased chance of being selected. There are several methods for selection: roulette wheel selection and its extensions, scaling techniques, tournament, elitist models, and ranking methods [53, 107]. In our case, the normalized geometric ranking method [81] is used since it only requires the evaluation function to map the solutions to a partially ordered set and it tends to eliminate chromosomes with extreme values, thus allowing for minimization and negativity. In normalized geometric ranking methods, Joines and Houck [81] define a probability of selection P_i for each individual as

$$P[\text{selecting the } i\text{th individual}] = q'(1-q)^{(r-1)}, \tag{9.1}$$

   where q represents the probability of selecting the best individual; r represents the rank of the individual, where 1 is the best; P is the population size; and $q' = q/(1-(1-q)^P)$.

3. Genetic operators:
   The operators are used to create new children based on the current generation in the population. Basically, there are two types of operators: crossover and mutation. Crossover takes two individuals and produces two new individuals while mutation alters one individual to produce a single new solution.

   In binary representations, the applications of these two types of operators are only binary mutation and simple crossover.

   In real-valued representations, the applications of these two types of operators have been developed by Michalewicz [107], they are uniform mutation, nonuniform mutation, multi-nonuniform mutation, boundary mutation, simple crossover, arithmetic crossover, and heuristic crossover [107].

   Uniform mutation randomly selects one variable and sets it equal to a uniform random number while boundary mutation randomly selects one variable and sets it equal to either its lower or upper bound.

   Nonuniform mutation randomly selects one variable and sets it equal to a nonuniform random number; according to [107], it is defined as follows: If ${s_x}^t = (x_1, x_2, x_3, \ldots, x_m)$ is a chromosome (t is the generation number) and the element x_j was selected for nonuniform mutation, the result is a vector ${s_x}^{t+1} = (x_1, x_2, x_3, \ldots, x'_j, \ldots, x_m)$, where

$$x'_j = x_j + \Delta(t, \mathrm{UB} - x_j), \quad \text{if a random digit is 0,} \tag{9.2}$$

$$x'_j = x_j - \Delta(t, x_j - \mathrm{LB}), \quad \text{if a random digit is 1,} \tag{9.3}$$

where UB and LB are the upper and lower bounds for the variable and $\Delta(t, y)$ is given by

$$\Delta(t, y) = y(1 - r^{(1-\frac{t}{G})^b}), \tag{9.4}$$

where r is a uniform random number between (0,1), G represents the maximum number of generations, t is the current generation, and b is a parameter determining the degree of dependency on the generation number.

4. Population size:
   The population size represents the number of individuals or chromosomes in the population. Usually, larger population sizes increase the amount of variation present in the initial population and it requires more fitness evaluations. If the population loses diversity, the population is said to have premature convergence and little exploration is being done. For longer chromosomes and challenging optimization problems, larger population sizes are needed to maintain diversity – higher diversity can also be achieved through higher mutation rates and uniform crossover – and hence better exploration. Usually, the population size is determined by the rule of thumb of 7–8 times the number of the optimization parameters.
5. Mutation rate:
   The mutation rate is defined as the percentage of the total number of genes in the population [51]; it determines the probability that a mutation will occur. Mutation is employed to give new information to the population and also prevents the population from becoming saturated with similar chromosomes (premature convergence). Large mutation rates increase the probability that good schemata will be destroyed, but increase population diversity. The best mutation rate is application dependent but for most applications it is between 0.001 and 0.1.
6. Crossover rate:
   The crossover rate (denoted by p_c) is defined as the ratio of the number of offspring produced in each generation to the population size, P [51]. This ratio controls the expected number $p_c \times P$ of chromosomes to undergo the crossover operation. The best crossover rate is application dependent but for most applications it is between 0.80 and 0.95.
7. Evaluation functions:
   Evaluation functions are subject to the minimal requirement that the function can map the population into a partially ordered set. In the present work, the sum of diagonal elements in stiffness/compliance matrix with relative weight factors for each direction is set as the evaluation function.

9.4 Single-Objective Optimization

9.4.1 *Objective of Global Stiffness*

In this research, the stiffness for certain mechanism configurations is expressed by a (6×6) matrix, as discussed before. The diagonal elements of the matrix are the mechanism's pure stiffness in each Cartesian direction. To obtain the optimal stiffness in each direction, one can write an objective function, (9.5), with stiffness element to maximize or write an objective function, (9.6), with compliance elements whose negative is to be maximized, i.e., maximize(val) where

$$\text{val} = \eta_1 K_{11} + \eta_2 K_{22} + \eta_3 K_{33} + \eta_4 K_{44} + \eta_5 K_{55} + \eta_6 K_{66} \tag{9.5}$$

or

$$\text{val} = -(\lambda_1 \kappa_{11} + \lambda_2 \kappa_{22} + \lambda_3 \kappa_{33} + \lambda_4 \kappa_{44} + \lambda_5 \kappa_{55} + \lambda_6 \kappa_{66}), \tag{9.6}$$

where, for $i = 1, \ldots, 6$, K_{ii} represents the diagonal elements of the mechanism's stiffness matrix, κ_{ii} represents the diagonal elements of mechanism's compliance matrix, η_i is the weight factor for each directional stiffness, which characterizes the priority of the stiffness in this direction, and λ_i is the weight factor for each directional compliance, which characterizes the priority of the compliance in this direction.

This would maximize/minimize the SUM of the diagonal elements. Although we could not maximize/minimize each diagonal element individually, we always can optimize each stiffness by distributing the weighting factors. Once the objective function is written, a search domain for each optimization variable (lengths, angles, etc.) should be specified to create an initial population. The limits of the search domain are set by a specified maximum number of generations or population convergence criteria, since the GAs will force much of the entire population to converge to a single solution.

For the optimization of the stiffness, a real-valued method is used combined with the selection, mutation, and crossover operators with their optional parameters used for all these types of parallel mechanism stiffness/compliance function optimization as shown in Table 9.1. The first optional parameter is the number of times to apply the operators for real-valued representation, G_m represents the maximum number of generations, and b is a parameter determining the degree of dependency on the generation number, we use 3 in our case [107]. The other optional parameters depend on the operators we are using. Since Matlab requires matrices to have the same length in all rows, many of the parameters are 0 indicating that they are really place holders only. In the following sections, we will describe it in more detail.

Table 9.1 Genetic algorithm parameters used for real-valued stiffness function optimization

Name	Parameters
Uniform mutation	[4 0 0]
Nonuniform mutation	[4 G_m b]
Multi-nonuniform mutation	[6 G_m b]
Boundary mutation	[4 0 0]
Simple crossover	[2 0]
Arithmetic crossover	[2 0]
Heuristic crossover	[2 3]
Normalized geometric selection	0.08

9.4.2 Spatial Six-Degree-of-Freedom Mechanism with Prismatic Actuators

The spatial 6-dof mechanism with prismatic actuators is shown in Figs. 5.2 and 5.3. In order to obtain the maximum global stiffness, five architecture and behavior parameters are used as optimization parameters, the vector of optimization variables is

$$\mathbf{s} = [R_p, R_b, z, T_p, T_b], \tag{9.7}$$

where R_p is the radius of the platform, R_b is the radius of the base, z is the height of the platform, T_p and T_b are the angles to determine the attachment points on the platform and on the base, and their bounds are

$$\begin{gathered} R_p \in [5, 10]\,\text{cm}, \; R_b \in [12, 22]\,\text{cm}, \\ z \in [45, 56]\,\text{cm}, \\ T_p \in [18, 26]^\circ, \; T_b \in [38, 48]^\circ. \end{gathered}$$

In this research work, the objective function of (9.5) is maximized where the following is assumed

$$\begin{aligned} \eta_i &= 1, \quad i = 1, \ldots, 6, \\ P &= 80, \\ G_{max} &= 100, \end{aligned}$$

where P is the population size and G_{max} is the maximum number of generations.

The genetic algorithm is implemented in Matlab to search for the best solutions. The results are given only for one case with $\phi = 0, \theta = 0$, and $\psi = 0$. Figures 9.2 and 9.3 show the evolution of the best individual and the optimal parameters for 40 generations, respectively. The architectural and behavior parameters found by the GA after 40 generations are

$$\mathbf{s} = [R_p, R_b, z, T_p, T_b] = [10, 12, 56, 18, 48]$$

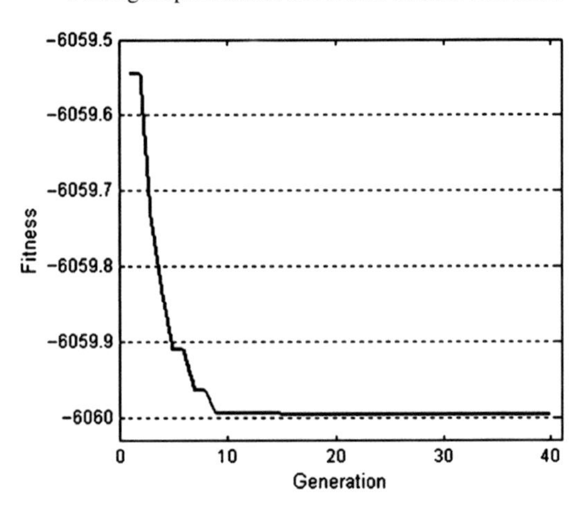

Fig. 9.2 The evolution of the performance of the 6-dof mechanism with prismatic actuators

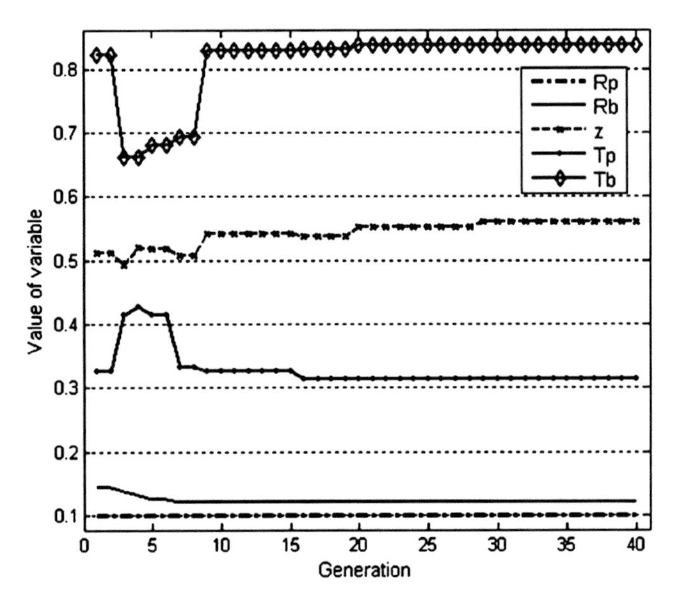

Fig. 9.3 The evolution of the geometrical parameters of the 6-dof mechanism with prismatic actuators

and the stiffness in each direction is

$$\begin{aligned}\mathbf{K} &= [K_x, K_y, K_z, K_{\theta_x}, K_{\theta_y}, K_{\theta_z}] \\ &= [34.1918, 34.1918, 5931.6164, 29.65808182, 29.65808182, 0.68092535].\end{aligned}$$

The sum of the stiffness is 6059.997.

Before optimization, the parameters for this mechanism were given as

$$\mathbf{s}' = [R_\mathrm{p}, R_\mathrm{b}, z, T_\mathrm{p}, T_\mathrm{b}] = [6, 15, 51, 22.34, 42.88]$$

and the stiffness in each direction was

$$\begin{aligned}\mathbf{K}' &= [K'_x, K'_y, K'_z, K'_{\theta_x}, K'_{\theta_y}, K'_{\theta_z}]\\ &= [102.968, 102.968, 5794.06, 10.4293, 10.4293, 0.222188]\end{aligned}$$

and the stiffness sum is 6021.08. Hence, after optimization, the stiffness sum is improved 1.01 times.

9.4.3 *Spatial Six-Degree-of-Freedom Mechanism with Revolute Actuators*

A spatial 6-dof mechanism with revolute actuators is represented in Fig. 6.2. The vertex distribution is the same as in Fig. 5.3. From Fig. 6.3, it is clear that the Cartesian stiffness is a monotonically increasing function of the link stiffness (for all the case with revolute actuators). Nevertheless, there exists a critical link stiffness, which has tiny effects on mechanism's Cartesian stiffness when it is larger than the critical link stiffness, therefore, for all mechanisms with revolute actuators, link stiffness are also included as optimization parameters. Seven optimization parameters are specified in this mechanism for maximizing the mechanism's global stiffness. The vector of optimization variables can be expressed as

$$\mathbf{s} = [R_\mathrm{p}, R_\mathrm{b}, z, T_\mathrm{p}, T_\mathrm{b}, l_1, l_2], \tag{9.8}$$

where R_p is the radius of the platform, R_b is the radius of the base, z is the height of the platform, T_p and T_b are the angles to determine the attachment points on the platform and on the base, l_1 and l_2 are the link lengths, and the bound for each parameter is

$$\begin{gathered}R_\mathrm{p} \in [5, 7]\,\mathrm{cm},\ R_\mathrm{b} \in [14, 16]\,\mathrm{cm},\\ z \in [66, 70]\,\mathrm{cm},\\ T_\mathrm{p} \in [20, 26]^\circ,\ T_\mathrm{b} \in [40, 45]^\circ,\\ l_1 \in [42, 48]\,\mathrm{cm},\ l_2 \in [32, 40]\,\mathrm{cm}.\end{gathered}$$

In this case, the objective function of (9.5) is maximized assuming

$$\begin{aligned}\eta_i &= 1, \quad i = 1, \ldots, 6,\\ P &= 80,\\ G_{\max} &= 40,\end{aligned}$$

where P is the population size and $G_{\max}$ is the maximum number of generations.

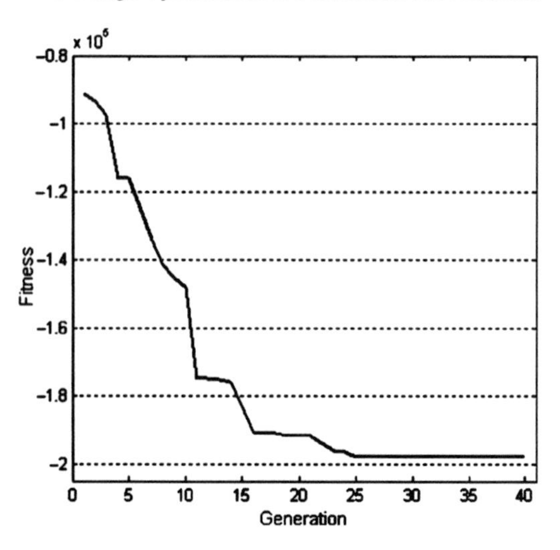

Fig. 9.4 The evolution of the performance of the 6-dof mechanism with revolute actuators

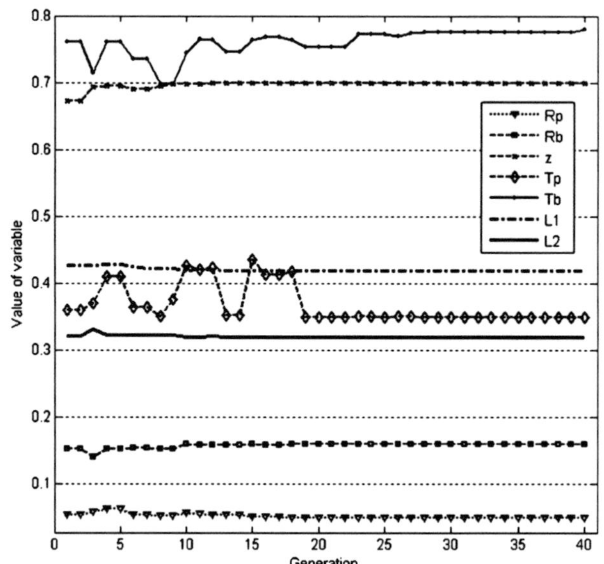

Fig. 9.5 The evolution of the geometrical parameters of the 6-dof mechanism with revolute actuators

A program based on the genetic algorithm is applied to search for the best solutions. The results are given only for one configuration with $\theta = 0, \phi = 0$, and $\psi = 0$. Figures 9.4 and 9.5 show that, after 40 generations, the track of the best individual and the optimal parameters converge to the final best solution. The optimal geometric and behavior parameters obtained by the GA after 40 generations are

$$\mathbf{s} = [R_\mathrm{p}, R_\mathrm{b}, z, T_\mathrm{p}, T_\mathrm{b}, l_1, l_2] = [5, 16, 70, 20, 45, 42, 32]$$

and the stiffness in each direction is

$$\begin{aligned}\mathbf{K} &= [K_x, K_y, K_z, K_{\theta_x}, K_{\theta_y}, K_{\theta_z}] \\ &= [18873.14, 18873.14, 159835.85, 199.79, 199.79, 41.04].\end{aligned}$$

The sum of the stiffness is 198022.766.

Initially, the geometric and behavior values were given for this mechanism as

$$\begin{aligned}\mathbf{s}' &= [R_\mathrm{p}, R_\mathrm{b}, z, T_\mathrm{p}, T_\mathrm{b}, l_1, l_2] \\ &= [6, 15, 68, 22.34, 42.883, 46, 36]\end{aligned}$$

and the stiffness in each direction was

$$\begin{aligned}\mathbf{K}' &= [K'_x, K'_y, K'_z, K'_{\theta_x}, K'_{\theta_y}, K'_{\theta_z}] \\ &= [7725, 7725, 21045, 37.8818, 37.8818, 43.0695].\end{aligned}$$

The stiffness sum is 36613.8. Therefore, after optimization, the stiffness sum is improved 5.4 times.

9.4.4 *Spatial Five-Degree-of-Freedom Mechanism with Prismatic Actuators*

The spatial 5-dof mechanism with prismatic actuators is shown in Figs. 5.6 and 5.7. In order to obtain the maximum global stiffness, the global compliance (since there are infinite terms among the diagonal stiffness elements) is minimized. However, it is clear that the Cartesian stiffness is a monotonically increasing function of the link and actuator stiffness (for all the case with prismatic actuators). Hence, the optimum solution always corresponds to the maximum link or actuator stiffness and these parameters are not included in the optimization variables. Seven parameters are specified as optimization parameters, they are

$$\mathbf{s} = [R_\mathrm{p}, R_\mathrm{b}, l_{61}, l_{62}, z, T_\mathrm{p}, T_\mathrm{b}], \tag{9.9}$$

where R_p is the radius of the platform; R_b is the radius of the base; l_{61} and l_{62} are the link length for the first and second link of the passive leg, respectively; z is the height of the platform; T_p and T_b are the angles to determine the attachment points on the platform and on the base; and their bounds are

$$\begin{gathered}R_\mathrm{p} \in [10, 14]\,\mathrm{cm},\, R_\mathrm{b} \in [20, 26]\,\mathrm{cm}, \\ l_{61} \in [52, 70]\,\mathrm{cm},\, l_{62} \in [52, 70]\,\mathrm{cm}, \\ z \in [66, 70]\,\mathrm{cm}, \\ T_\mathrm{p} \in [18, 26]^\circ,\, T_\mathrm{b} \in [38, 48]^\circ.\end{gathered}$$

In this work, the objective function of (9.6) is minimized assuming

$$\begin{aligned}\lambda_i &= 1, \quad i = 1,\ldots,6,\\ P &= 80,\\ G_{\max} &= 40.\end{aligned}$$

Results are given here only for the case with $\theta_{65} = -\pi$ and $\theta_{66} = 2\pi/3$. Figures 9.6 and 9.7 show the evolution of the best individual and the optimal parameters for 40 generations, respectively. The architectural and behavior parameters found by the GA after 40 generations are

$$\begin{aligned}\mathbf{s} &= [R_{\mathrm{p}}, R_{\mathrm{b}}, l_{61}, l_{62}, z, T_{\mathrm{p}}, T_{\mathrm{b}}]\\ &= [14, 21.2, 52, 70, 66, 18, 48]\end{aligned}$$

and the compliances in each direction are

$$\begin{aligned}\kappa &= [\kappa_{\theta_x}, \kappa_{\theta_y}, \kappa_{\theta_z}, \kappa_x, \kappa_y, \kappa_z]\\ &= [0.03687, 0.03113, 0.03646, 0.03962, 0.01657, 2.46\times 10^{-4}].\end{aligned}$$

The sum of the compliances is 0.16.

Before optimization, the parameter values of the mechanism were given as

$$\begin{aligned}\mathbf{s}' &= [R_{\mathrm{p}}, R_{\mathrm{b}}, l_{61}, l_{62}, z, T_{\mathrm{p}}, T_{\mathrm{b}}]\\ &= [12, 22, 68, 68, 68, 22.34, 42.883]\end{aligned}$$

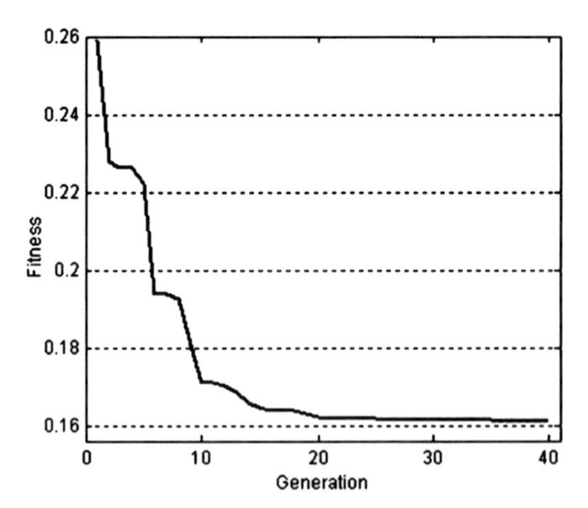

Fig. 9.6 The evolution of the performance of the 5-dof mechanism with prismatic actuators

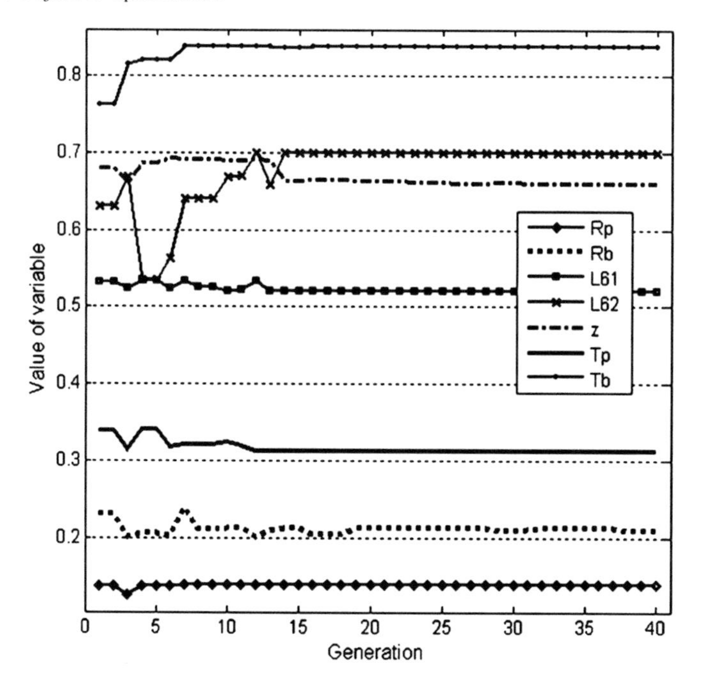

Fig. 9.7 The evolution of the geometrical parameters of the 5-dof mechanism with prismatic actuators

and the compliances in each direction were

$$\begin{aligned}\kappa' &= [\kappa'_{\theta_x}, \kappa'_{\theta_y}, \kappa'_{\theta_z}, \kappa'_x, \kappa'_y, \kappa'_z] \\ &= [0.08627, 0.0981, 0.2588, 0.07342, 0.030325, 2.55 \times 10^{-4}].\end{aligned}$$

The compliance sum is 0.54714. After optimization, the compliance sum is improved 3.4 times.

9.4.5 *Spatial Five-Degree-of-Freedom Mechanism with Revolute Actuators*

The schematic representation of this type mechanism and its vertex distribution is shown in Figs. 6.5 and 6.6, respectively. Twelve architecture and behavior parameters are specified as optimization parameters to minimize the compliances. They can be represented as a vector of **s**

$$\mathbf{s} = [R_p, R_b, l_{61}, l_{62}, l_1, l_2, z, T_p, T_b], \qquad (9.10)$$

where R_p is the radius of the platform; R_b is the radius of the base; l_{61} and l_{62} are the link lengths for the first and second link of the passive leg, respectively; l_1 and l_2 are the link lengths for the first and second link of the each actuated leg, respectively; z is the height of the platform; T_p and T_b are the angles to determine the attachment points on the platform and on the base; and the bound of each optimization parameter is

$$\begin{gathered} R_p \in [5, 7]\,\text{cm}, R_b \in [14, 18]\,\text{cm}, \\ l_{61} \in [67, 70]\,\text{cm}, l_{62} \in [67, 70]\,\text{cm}, \\ l_1 \in [33, 35]\,\text{cm}, l_2 \in [45, 47]\,\text{cm}, \\ z \in [66, 70]\,\text{cm}, T_p \in [18, 30]^\circ, T_b \in [38, 50]^\circ. \end{gathered}$$

For this mechanism, the objective function of (9.6) is minimized with

$$\begin{aligned} \lambda_i &= 1, \quad i = 1, \ldots, 6, \\ P &= 80, \\ G_{\max} &= 40. \end{aligned}$$

Results are given here for the case with $\theta_{65} = -\pi$ and $\theta_{66} = 2\pi/3$. Figures 9.8 and 9.9 show the evolution of the best individual and the optimal parameters for 40 generations, respectively. The mechanism's geometric and behavior parameters found by the GA after 40 generations are

$$\begin{aligned} \mathbf{s} &= [R_p, R_b, l_{61}, l_{62}, l_1, l_2, z, T_p, T_b] \\ &= [7, 18, 69.052, 67.179, 33, 45, 70, 18.92, 50.03] \end{aligned}$$

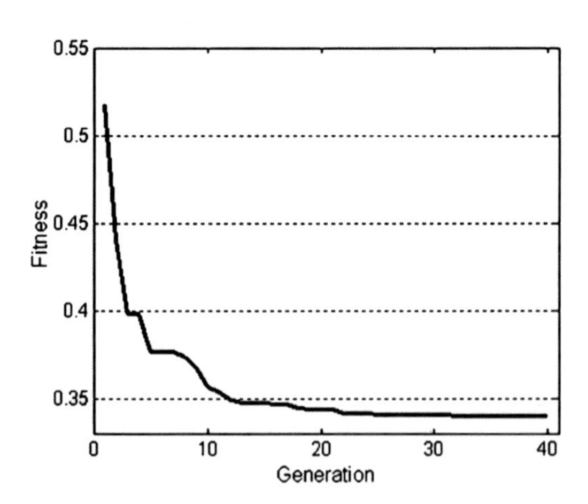

Fig. 9.8 The evolution of the performance of the 5-dof mechanism with revolute actuators

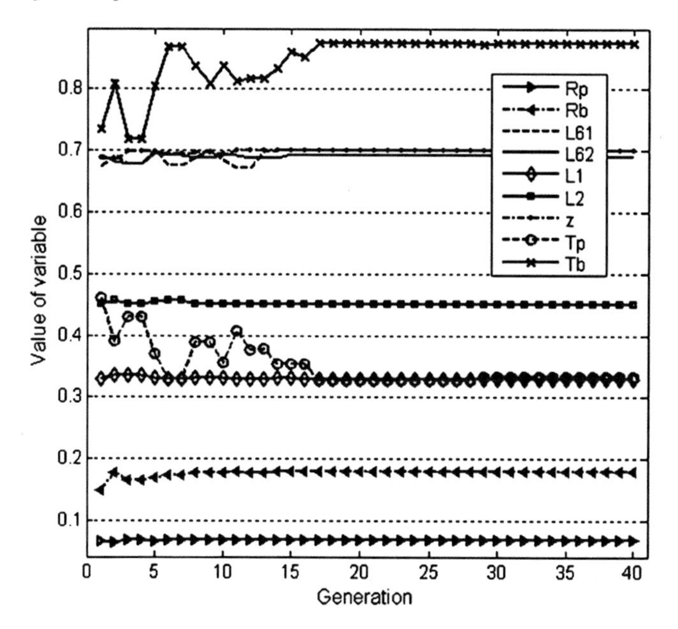

Fig. 9.9 The evolution of the geometrical parameters of the 5-dof mechanism with revolute actuators

and the compliances in each direction are

$$\begin{aligned}\kappa &= [\kappa_{\theta_x}, \kappa_{\theta_y}, \kappa_{\theta_z}, \kappa_x, \kappa_y, \kappa_z] \\ &= [7.77 \times 10^{-2}, 0.10345, 0.24256, 1.116 \times 10^{-3}, 1.87 \times 10^{-3}, 2.67 \times 10^{-4}].\end{aligned}$$

The sum of the compliances is 0.33977.

The initial guess of the geometric and structure behavior parameters of the mechanism was given as

$$\begin{aligned}\mathbf{s}' &= [R_{\mathrm{p}}, R_{\mathrm{b}}, l_{61}, l_{62}, l_1, l_2, z, T_{\mathrm{p}}, T_{\mathrm{b}}] \\ &= [6, 15, 68, 68, 34, 46, 68, 22.34, 42.883]\end{aligned}$$

and the compliances in each direction were

$$\begin{aligned}\kappa' &= [\kappa'_{\theta_x}, \kappa'_{\theta_y}, \kappa'_{\theta_z}, \kappa'_x, \kappa'_y, \kappa'_z] \\ &= [0.1244, 0.2327, 0.3732, 0.001, 0.002464, 3.589 \times 10^{-4}].\end{aligned}$$

The compliance sum is 0.734195. Hence after optimization, the compliance sum is improved 2.16 times.

9.4.6 Spatial Four-Degree-of-Freedom Mechanism with Prismatic Actuators

Figure 5.8 shows the spatial 4-dof mechanism with prismatic actuators. For this mechanism, the optimization parameters are

$$\mathbf{s} = [R_\mathrm{p}, R_\mathrm{b}, l_{51}, l_{52}, z, T_\mathrm{p}, T_\mathrm{b}], \tag{9.11}$$

where R_p is the radius of the platform; R_b is the radius of the base; l_{51} and l_{52} are the link lengths for the 1st and 2nd link of the passive leg, respectively; z is the height of the platform; T_p and T_b are the angles to determine the attachment points on the platform and on the base, and their bounds are

$$R_\mathrm{p} \in [10, 14]\,\mathrm{cm},\, R_\mathrm{b} \in [20, 26]\,\mathrm{cm},$$
$$l_{51} \in [52, 70]\,\mathrm{cm},\, l_{52} \in [52, 70]\,\mathrm{cm},$$
$$z \in [66, 70]\,\mathrm{cm},\, T_\mathrm{p} \in [25, 35]^\circ,\, T_b \in [55, 65]^\circ,$$

Again, the compliances are minimized as above.

Results are given here only for one case with $\theta_{55} = -\pi/3$ and $\theta_{56} = 2\pi/3$. Figures 9.10 and 9.11 show the evolution of the best individual and the optimal parameters for 40 generations, respectively. The geometric and behavior parameters found by the GA after 40 generations are

$$\begin{aligned}\mathbf{s} &= [R_\mathrm{p}, R_\mathrm{b}, l_{51}, l_{52}, z, T_\mathrm{p}, T_\mathrm{b}] \\ &= [14, 26, 70, 55, 66, 35, 55]\end{aligned}$$

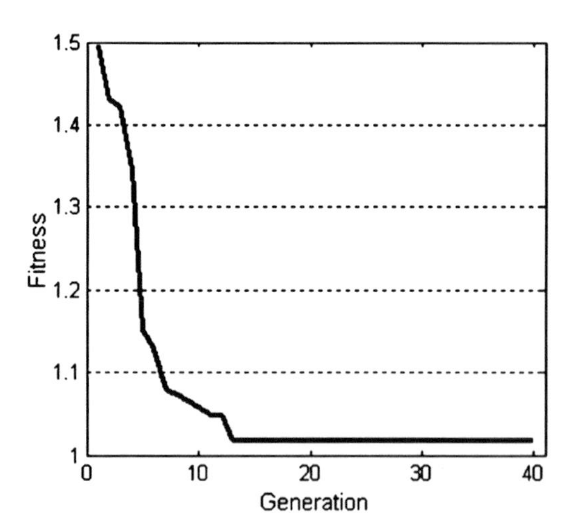

Fig. 9.10 The evolution of the performance of the 4-dof mechanism with prismatic actuators

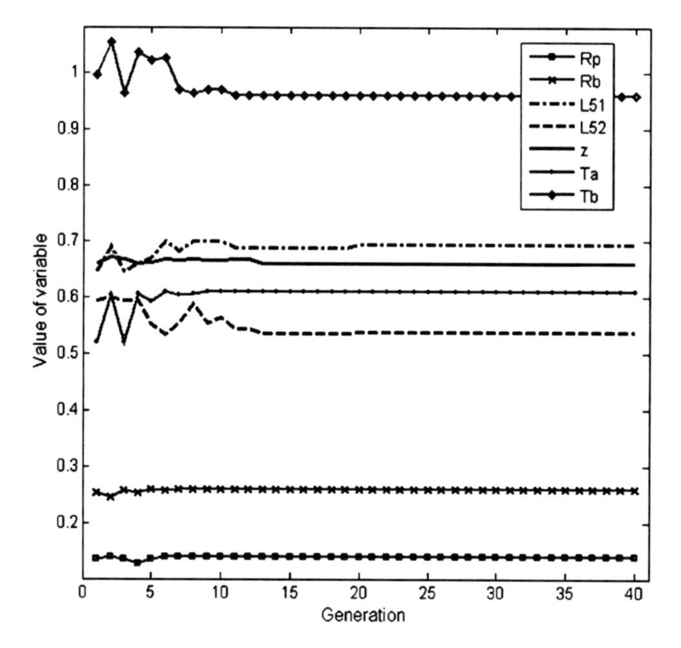

Fig. 9.11 The evolution of the geometrical parameters of the 4-dof mechanism with prismatic actuators

and the compliances in each direction are

$$\begin{aligned}\kappa &= [\kappa_{\theta_x}, \kappa_{\theta_y}, \kappa_{\theta_z}, \kappa_x, \kappa_y, \kappa_z] \\ &= [0.12, 0.5742, 3.747 \times 10^{-3}, 0.3165, 5.006 \times 10^{-11}, 3.345 \times 10^{-3}].\end{aligned}$$

The sum of the compliances is 1.017897.

Initially, the parameters for this mechanism were given as

$$\begin{aligned}\mathbf{s}' &= [R_\mathrm{p}, R_\mathrm{b}, l_{51}, l_{52}, z, T_\mathrm{p}, T_\mathrm{b}] \\ &= [12, 22, 68, 68, 68, 30, 60]\end{aligned}$$

and the compliances in each direction were

$$\begin{aligned}\kappa' &= [\kappa'_{\theta_x}, \kappa'_{\theta_y}, \kappa'_{\theta_z}, \kappa'_x, \kappa'_y, \kappa'_z] \\ &= [0.5164, 1.4046, 1.5 \times 10^{-10}, 0.9087, 5.78 \times 10^{-11}, 0.011139].\end{aligned}$$

The compliance sum is 2.84085. Therefore, after optimization the compliance sum has been improved 2.8 times.

9.4.7 Spatial Four-Degree-of-Freedom Mechanism with Revolute Actuators

A spatial 4-dof mechanism with revolute actuators is shown in Fig. 6.9. The parameters are

$$\mathbf{s} = [R_p, R_b, l_{51}, l_{52}, L_1, L_2, z, T_p, T_b], \tag{9.12}$$

where R_p is the radius of the platform; R_b is the radius of the base; l_{51} and l_{52} are the link lengths for the first and second link of the passive leg, respectively; L_1 and L_2 are the link lengths for the first and second link of the actuated leg, respectively; z is the height of the platform; T_p and T_b are the angles to determine the attachment points on the platform and on the base; and their bounds are

$$\begin{gathered} R_p \in [5, 7]\,\text{cm},\ R_b \in [14, 16]\,\text{cm}, \\ l_{51} \in [67, 69]\,\text{cm},\ l_{52} \in [67, 69]\,\text{cm}, \\ L_1 \in [33, 35]\,\text{cm},\ L_2 \in [45, 47]\,\text{cm}, \\ z \in [66, 70]\,\text{cm}, \\ T_p \in [25, 35]^\circ,\ T_b \in [55, 65]^\circ, \end{gathered}$$

Results are given here for one case with $\theta_{55} = -\pi/3$ and $\theta_{56} = 2\pi/3$. Figures 9.12 and 9.13 show the evolution of the best individual and the optimal parameters for 40 generations, respectively. The optimum geometric and behavior parameters for this configuration are

$$\begin{aligned} \mathbf{s} &= [R_p, R_b, l_{51}, l_{52}, L_1, L_2, z, T_p, T_b] \\ &= [7, 16, 67, 69, 35, 47, 66, 35, 55] \end{aligned}$$

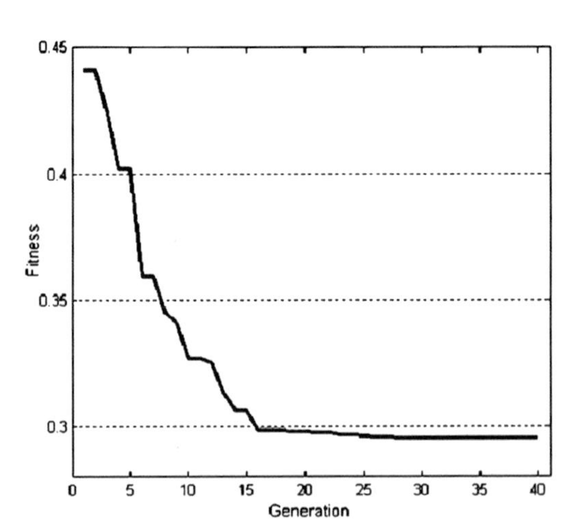

Fig. 9.12 The evolution of the performance of the 4-dof mechanism with revolute actuators

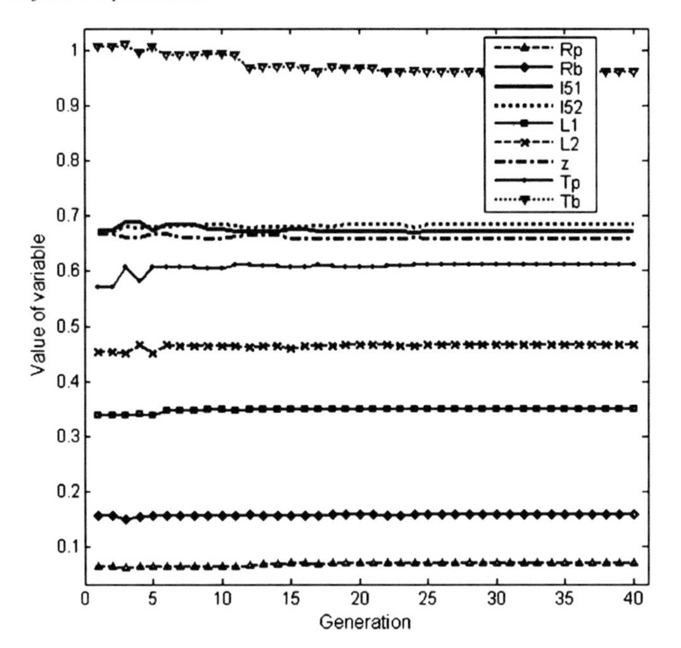

Fig. 9.13 The evolution of the geometrical parameters of the 4-dof mechanism with revolute actuators

and the compliances in each direction are

$$\begin{aligned}\kappa &= [\kappa_{\theta_x}, \kappa_{\theta_y}, \kappa_{\theta_z}, \kappa_x, \kappa_y, \kappa_z] \\ &= [0.28857, 0.019376, 8.66 \times 10^{-5}, 9.39 \times 10^{-4}, 5.751 \times 10^{-11}, 3.646 \times 10^{-5}].\end{aligned}$$

The sum of the compliances is 0.29518.

The initial guess for this mechanism was

$$\begin{aligned}\mathbf{s}' &= [R_\mathrm{p}, R_\mathrm{b}, l_{51}, l_{52}, L_1, L_2, z, T_\mathrm{p}, T_\mathrm{b}] \\ &= [6, 15, 68, 68, 34, 46, 68, 30, 60]\end{aligned}$$

and the compliances in each direction were

$$\begin{aligned}\boldsymbol{\kappa}' &= [\kappa'_{\theta_x}, \kappa'_{\theta_y}, \kappa'_{\theta_z}, \kappa'_x, \kappa'_y, \kappa'_z] \\ &= [1.2807, 0.0628078, 0, 0.00276278, 0, 0.00003838].\end{aligned}$$

The compliance sum is 1.3463. Hence, after optimization, the total compliance is improved 4.56 times.

9.4.8 Spatial Three-Degree-of-Freedom Mechanism with Prismatic Actuators

The spatial 3-dof mechanism with prismatic actuators is shown in Fig. 5.9. The parameters are

$$\mathbf{s} = [R_\mathrm{p}, R_\mathrm{b}, z], \tag{9.13}$$

where R_p is the radius of the platform, R_b is the radius of the base, z is the height of the platform, and their bounds are set as

$$R_\mathrm{p} \in [5, 10]\,\mathrm{cm},\ R_\mathrm{b} \in [12, 14]\,\mathrm{cm},$$
$$z \in [66, 70]\,\mathrm{cm}.$$

Here only the case with $\theta_{45} = \pi/2$ and $\theta_{46} = 0$ is discussed. Figures 9.14 and 9.15 show the evolution of the best individual and the optimal parameters for 40 generations, respectively. After 40 generations, the optimal geometric and behavior parameters found by the GA are

$$\begin{aligned}\mathbf{s} &= [R_\mathrm{p}, R_\mathrm{b}, z]\\ &= [10, 12, 70]\end{aligned}$$

and the compliances in each direction are

$$\begin{aligned}\kappa &= [\kappa_{\theta_x}, \kappa_{\theta_y}, \kappa_{\theta_z}, \kappa_x, \kappa_y, \kappa_z]\\ &= [6.8355 \times 10^{-2}, 6.8355 \times 10^{-2}, 0, 0, 0, 3.4177 \times 10^{-4}].\end{aligned}$$

The sum of the compliances is 0.137.

The initial geometric and behavior values for this mechanism were given as

$$\begin{aligned}\mathbf{s}' &= [R_\mathrm{p}, R_\mathrm{b}, z]\\ &= [6, 15, 68]\end{aligned}$$

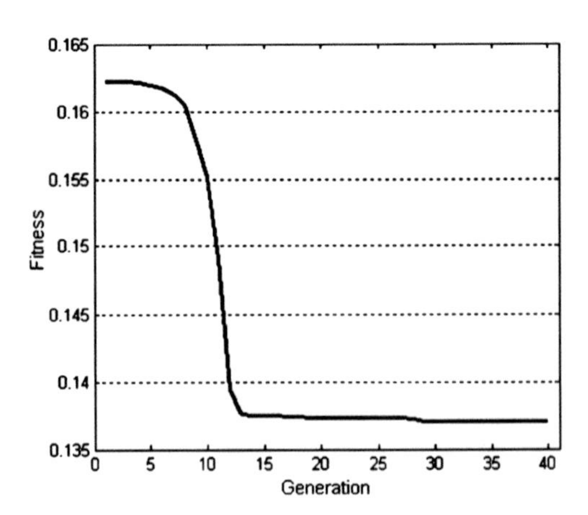

Fig. 9.14 The evolution of the performance of the 3-dof mechanism with prismatic actuators

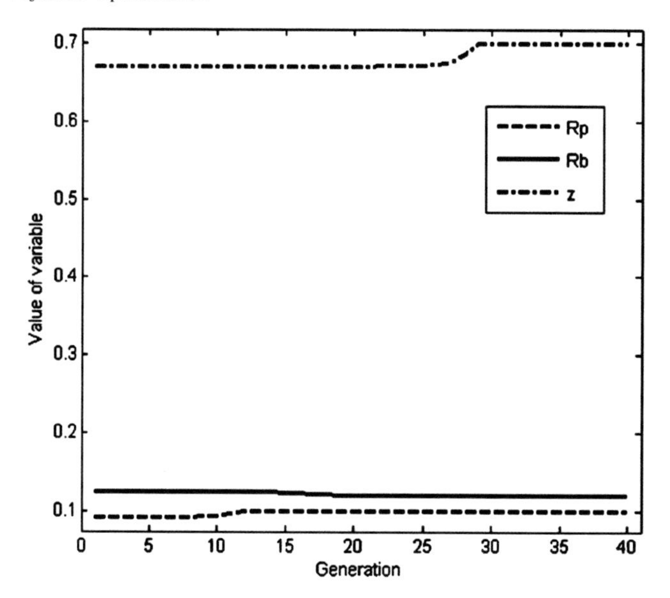

Fig. 9.15 The evolution of the geometrical parameters of the 3-dof mechanism with prismatic actuators

and the compliances in each direction were

$$\begin{aligned}\kappa' &= [\kappa'_{\theta_x}, \kappa'_{\theta_y}, \kappa'_{\theta_z}, \kappa'_x, \kappa'_y, \kappa'_z] \\ &= [0.192, 0.192, 0, 0, 0, 3.4566 \times 10^{-4}].\end{aligned}$$

The compliance sum is 0.3844. Therefore, after optimization, the total compliance is improved 2.81 times which is a minor gain.

9.4.9 Spatial Three-Degree-of-Freedom Mechanism with Revolute Actuators

The spatial 3-dof mechanism with revolute actuators is shown in Fig. 6.10. The parameters are

$$\mathbf{s} = [R_p, R_b, l_1, l_2, z], \tag{9.14}$$

where R_p is the radius of the platform, R_b is the radius of the base, l_1 and l_2 are the link length, z is the height of the platform, and their bounds are

$$\begin{gathered}R_p \in [5, 7]\,\text{cm},\, R_b \in [14, 16]\,\text{cm}, \\ l_1 \in [33, 35]\,\text{cm},\, l_2 \in [45, 47]\,\text{cm}, \\ z \in [66, 70]\,\text{cm}.\end{gathered}$$

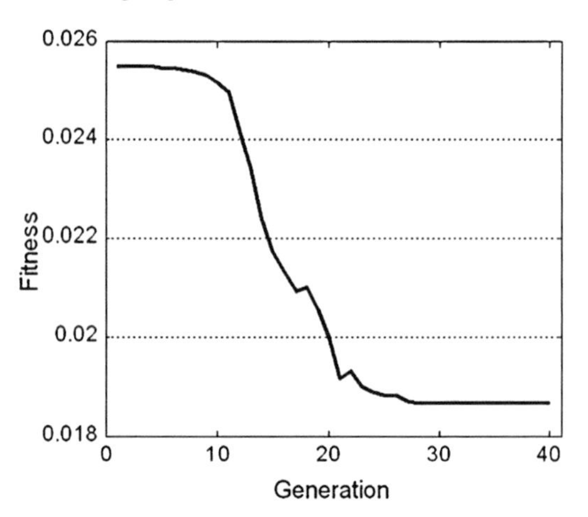

Fig. 9.16 The evolution of the performance of the 3-dof mechanism with revolute actuators

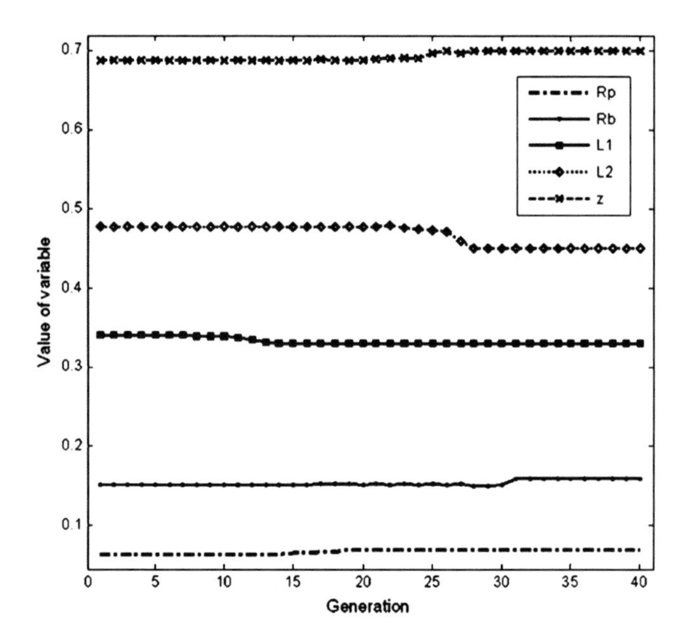

Fig. 9.17 The evolution of the geometrical parameters of the 3-dof mechanism with revolute actuators

Here only one case with $\theta_{45} = \pi/2$ and $\theta_{46} = 0$ is analyzed. Figures 9.16 and 9.17 show the evolution of the best individual and the optimal parameters for 40 generations, respectively. After running the program for 40 generations, the optimal architectural and behavior parameters can be found as

$$
\begin{aligned}
\mathbf{s} &= [R_{\mathrm{p}}, R_{\mathrm{b}}, l_1, l_2, z] \\
&= [7, 16, 33, 45, 70]
\end{aligned}
$$

and the compliances in each direction are

$$\begin{aligned}\boldsymbol{\kappa} &= [\kappa_{\theta_x}, \kappa_{\theta_y}, \kappa_{\theta_z}, \kappa_x, \kappa_y, \kappa_z] \\ &= [1.0782 \times 10^{-2}, 1.0782 \times 10^{-2}, 0, 0, 0, 2.64 \times 10^{-5}].\end{aligned}$$

The sum of the compliances is 0.018659.

Before optimization, a series of parameters were guessed as

$$\begin{aligned}\mathbf{s}' &= [R_\mathrm{p}, R_\mathrm{b}, l_1, l_2, z] \\ &= [6, 15, 34, 46, 68]\end{aligned}$$

and the compliance in each direction can be computed as

$$\begin{aligned}\boldsymbol{\kappa}' &= [\kappa'_{\theta_x}, \kappa'_{\theta_y}, \kappa'_{\theta_z}, \kappa'_x, \kappa'_y, \kappa'_z] \\ &= [2.12264 \times 10^{-2}, 2.12264 \times 10^{-2}, 0, 0, 0, 3.82 \times 10^{-5}].\end{aligned}$$

The compliance sum is 0.04249. Hence, after optimization, the total compliances is improved 2.28 times.

9.4.10 The Tricept Machine Tool Family

The schematic representation of the Tricept machine tool and the geometry of the joint distribution both on the base and the platform are shown in Figs. 9.18 and 9.19, respectively. The vector of optimization variables is therefore

$$\mathbf{s} = [R_\mathrm{p}, R_\mathrm{b}, z], \tag{9.15}$$

where R_p is the radius of the platform, R_b is the radius of the base, z is the height of the platform, and their bounds are specified based on the dimensions of the Tricept machine tool

$$\begin{aligned}R_\mathrm{p} \in [200, 300]\,\mathrm{mm},\ \ R_\mathrm{b} \in [400, 600]\,\mathrm{mm}, \\ z \in [900, 1500]\,\mathrm{mm}.\end{aligned}$$

The case with $\theta_{41} = \pi/2$ and $\theta_{42} = 0$ is discussed here. Figures 9.20 and 9.21 show the evolution of the best individual and the optimal parameters for 40 generations, respectively. The optimal architectural and behavior parameters found by the GA after 40 generations are

$$\mathbf{s} = [R_\mathrm{p}, R_\mathrm{b}, z] = [300, 600, 900]$$

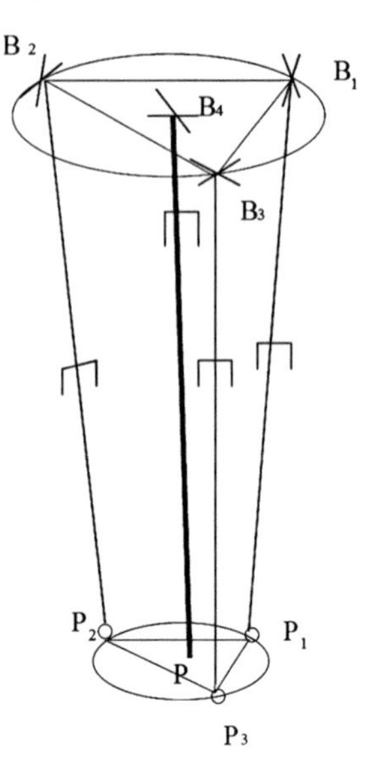

Fig. 9.18 Schematic representation of the Tricept machine tool

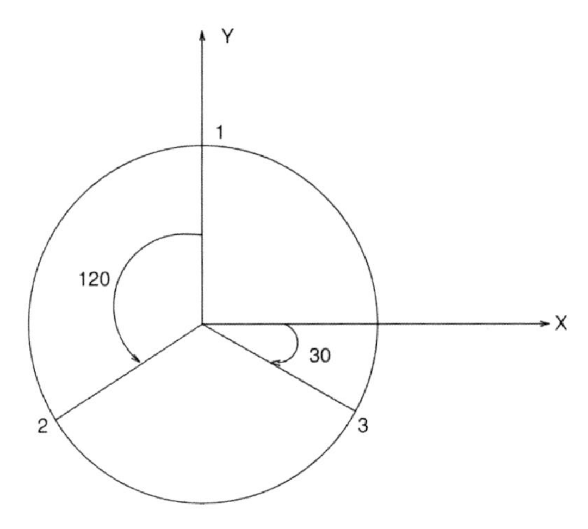

Fig. 9.19 Position of the attachment points on the base and platform

and the compliances in each direction are

$$\begin{aligned}\kappa &= \left[\kappa_{\theta_x}, \kappa_{\theta_y}, \kappa_{\theta_z}, \kappa_x, \kappa_y, \kappa_z\right] \\ &= [2.0576 \times 10^{-3}, 2.0576 \times 10^{-3}, 0, 1.667 \times 10^{-3}, 1.667 \times 10^{-3}, 3.703 \times 10^{-4}].\end{aligned}$$

The sum of the compliances is 0.0078189. Before optimization, the dimensions of the Tricept machine tool provided by Neos Robotics AB were

$$\mathbf{s}' = [R_\mathrm{p}, R_\mathrm{b}, z] = [225, 500, 1300]$$

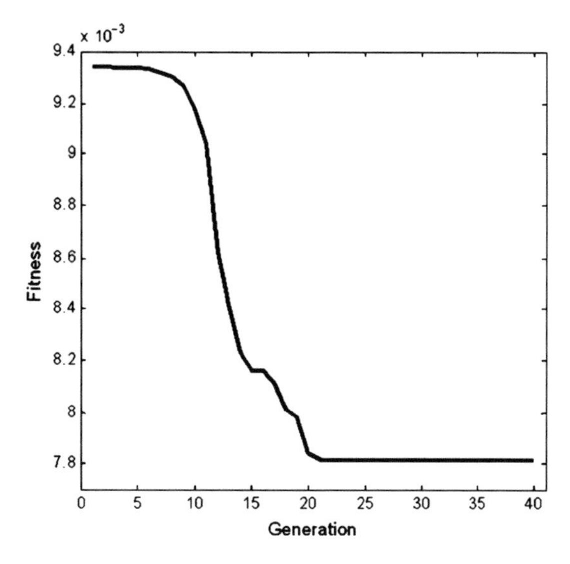

Fig. 9.20 The evolution of the performance of the Tricept machine tool

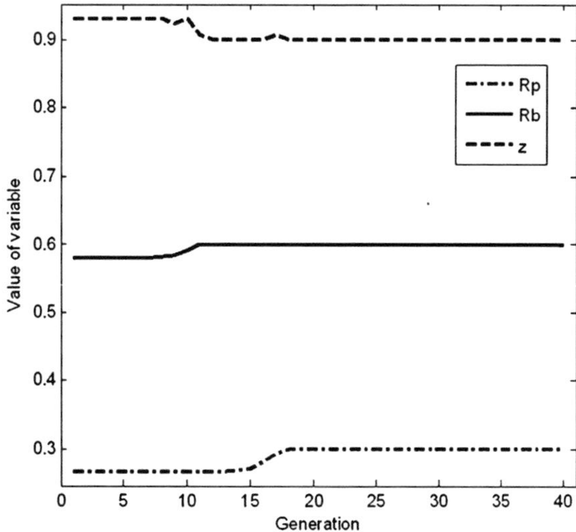

Fig. 9.21 The evolution of the geometrical parameters of the Tricept machine tool

and the compliances in each direction were

$$\begin{aligned}\kappa' &= [\kappa'_{\theta_x}, \kappa'_{\theta_y}, \kappa'_{\theta_z}, \kappa'_x, \kappa'_y, \kappa'_z] \\ &= [2.786 \times 10^{-3}, 2.786 \times 10^{-3}, 0, 4.708 \times 10^{-3}, 4.708 \times 10^{-3}, 3.4825 \times 10^{-4}].\end{aligned}$$

The sum of the compliances is 0.0153369. Hence, after optimization, the sum of the compliances is improved by a factor of 1.96 just by slightly enlarging the radius of the base and the platform.

9.5 Multiobjective Optimization

9.5.1 *Case Study 1: Three Degrees of Freedom Parallel Manipulator – Two Translations and One Rotation*

9.5.1.1 Structure Description

The new 3-dof parallel manipulator is composed of a base structure, a moving platform, and three legs connecting the base and platform. Among those three legs, two of them are in same plane and consist of identical planar four bar parallelograms as chains connected to the moving platform by revolute joints, while the third leg is one rectangular bar connected to the moving platform by a spatial joint. There is one revolute joint on the top end of each leg, and the revolute joint is linked to the base by an active prismatic joint.

The CAD model of the 3-dof parallel mechanism is shown in Fig. 9.22.

The dof for a closed-loop kinematic chain can be determined using the Chebychev–Grübler–Kutzbach formula [170]:

$$l = d(n - g - 1) + \sum_{i=1}^{g} f_i, \tag{9.16}$$

where l is the degree of freedom of the kinematic chain; d the degree of freedom of each unconstrained individual body (6 for the spatial case, 3 for the planar case);

Fig. 9.22 CAD modeling of 3-dof parallel manipulator

n the number of rigid bodies or links in the chain; g the number of joints; and f_i is the number of degrees of freedom allowed by the ith joint.

With eq. (9.16), the degree of freedom of the proposed parallel manipulator is

$$M = 6(8 - 9 - 1) + (5 + 5 + 5) = 3. \tag{9.17}$$

In Fig. 9.22, the parallelograms play the role of improving the kinematics performance and the leg stiffness can be increased largely [21]. In regard to the types of actuated joints, they can be either revolute or prismatic. Since the prismatic joints can easily achieve high accuracy and heavy loads, the majority of the 3-dof parallel mechanism in reality use actuated prismatic joints.

The output link of a planar parallelogram mechanism will remain in a fixed orientation with respect to its input link, and the parallelogram can ensure the desired output, in terms of translation and rotation. The advantages of the proposed parallel manipulator are as follows:

1. The use of the parallelogram structure can greatly increase the stiffness of the legs.
2. Two identical chains offer good symmetry.
3. The joint which connects the third leg and the moving platform gives the rotation about y-axis with respect to reference frame attached to the end-effecter.

A kinematics model of the manipulator is shown in Fig. 9.23. The vertices of the moving platform are p_i $(i = 1, 2, 3)$, and the vertices of the base are b_i $(i = 1, 2, 3)$.

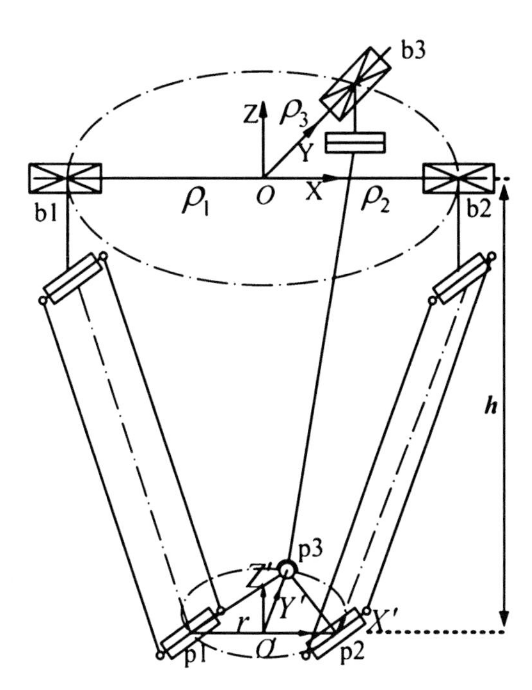

Fig. 9.23 Schematic representation

A global reference system $O : O - xyz$ is located at the point of intersection b_1b_2 and Ob_3. Another reference system, called the moving frame $O' : O' - x'y'z'$, is located at the center of p_1p_2 on the moving platform. The given position and orientation of the end-effecter (the moving platform) is specified by its three independent motions: y, z pure translations and ϕ pure rotation about y-axis. The position is given by the position vectors $(\mathbf{O}')_O$ and the orientation is given by rotation matrix $\mathbf{Q}$ as follows:

$$(\mathbf{O}')_O = (x\ y\ z)^{\mathrm{T}}, \tag{9.18}$$

where $x = 0$ and

$$\mathbf{Q} = \begin{bmatrix} \cos\phi & 0 & \sin\phi \\ 0 & 1 & 0 \\ -\sin\phi & 0 & \cos\phi \end{bmatrix}, \tag{9.19}$$

where angle ϕ is the rotation about y-axis. The coordinates of the point p_i in reference system $(\mathbf{O}')$ can be described by the vector $(\mathbf{p}_i)_{O'}$ $(i = 1, 2, 3)$

$$(\mathbf{p}_1)_{O'} = \begin{pmatrix} -r \\ 0 \\ 0 \end{pmatrix}, \ (\mathbf{p}_2)_{O'} = \begin{pmatrix} r \\ 0 \\ 0 \end{pmatrix}, \ (\mathbf{p}_3)_{O'} = \begin{pmatrix} 0 \\ r \\ 0 \end{pmatrix}. \tag{9.20}$$

The vectors $(\mathbf{b}_i)_O$ $(i = 1, 2, 3)$ in frame $O : O - xyz$ will be defined as position vectors of joints:

$$(\mathbf{b}_1) = \begin{pmatrix} -\rho_1 \\ 0 \\ 0 \end{pmatrix}, \ (\mathbf{b}_2) = \begin{pmatrix} \rho_2 \\ 0 \\ 0 \end{pmatrix}, \ (\mathbf{b}_3) = \begin{pmatrix} 0 \\ \rho_3 \\ 0 \end{pmatrix}. \tag{9.21}$$

The vector $(p_i)_O$ $(i = 1, 2, 3)$ in frame $O : O - xyz$ can be written as

$$(\mathbf{p}_i)_O = \mathbf{Q}(\mathbf{p}_i)_{O'} + (\mathbf{O}')_O. \tag{9.22}$$

That is

$$(\mathbf{p}_1)_O = \begin{bmatrix} \cos\phi & 0 & \sin\phi \\ 0 & 1 & 0 \\ -\sin\phi & 0 & \cos\phi \end{bmatrix} \begin{pmatrix} -r \\ 0 \\ 0 \end{pmatrix} + \begin{pmatrix} 0 \\ y \\ z \end{pmatrix} = \begin{pmatrix} -r\cos\phi \\ y \\ r\sin\phi + z \end{pmatrix}, \tag{9.23}$$

$$(\mathbf{p}_2)_O = \begin{bmatrix} \cos\phi & 0 & \sin\phi \\ 0 & 1 & 0 \\ -\sin\phi & 0 & \cos\phi \end{bmatrix} \begin{pmatrix} r \\ 0 \\ 0 \end{pmatrix} + \begin{pmatrix} 0 \\ y \\ z \end{pmatrix} = \begin{pmatrix} r\cos\phi \\ y \\ -r\sin\phi + z \end{pmatrix}, \tag{9.24}$$

$$(\mathbf{p}_3)_O = \begin{bmatrix} \cos\phi & 0 & \sin\phi \\ 0 & 1 & 0 \\ -\sin\phi & 0 & \cos\phi \end{bmatrix} \begin{pmatrix} 0 \\ r \\ 0 \end{pmatrix} + \begin{pmatrix} 0 \\ y \\ z \end{pmatrix} = \begin{pmatrix} 0 \\ r + y \\ z \end{pmatrix}. \tag{9.25}$$

The inverse kinematics of the manipulator can be solved by applying the following constraint equation:

$$\|\mathbf{p}_i - \mathbf{b}_i\| = L. \tag{9.26}$$

Hence, one can obtain the required actuator inputs:

$$\rho_1 = \sqrt{L^2 - y^2 - (z + r\sin\phi)^2} + r\cos\phi, \tag{9.27}$$

$$\rho_2 = \sqrt{L^2 - y^2 - (z - r\sin\phi)^2} + r\cos\phi, \tag{9.28}$$

$$\rho_3 = \sqrt{L^2 - z^2} + y + r. \tag{9.29}$$

Equations (9.27), (9.28), and (9.29) can be differentiated with respect to time to obtain the velocity equations,

$$(\rho_1 - r\cos\phi)\dot{\rho}_1 + y\dot{y} + (z + r\sin\phi)\dot{z} + r(\rho_1\sin\phi + z\cos\phi)\dot{\phi} = 0, \tag{9.30}$$

$$(\rho_2 - r\cos\phi)\dot{\rho}_2 + y\dot{y} + (z - r\sin\phi)\dot{z} + r(\rho_2\sin\phi - z\cos\phi)\dot{\phi} = 0, \tag{9.31}$$

$$[\rho_3 - (y + r)]\dot{\rho} + [\rho_3 - (y + r)]\dot{y} + z\dot{z} = 0. \tag{9.32}$$

Rearranging (9.30), (9.31), and (9.32) we have

$$\mathbf{A}\dot{\boldsymbol{\rho}} = \mathbf{B}\dot{\mathbf{p}}, \tag{9.33}$$

where $\dot{\boldsymbol{\rho}}$ is the vector of input velocities defined as

$$\dot{\boldsymbol{\rho}} = (\dot{\rho}_1, \dot{\rho}_2, \dot{\rho}_3)^{\mathrm{T}} \tag{9.34}$$

and $\dot{\mathbf{p}}$ is the vector of output velocities defined as

$$\dot{\mathbf{p}} = (\dot{y}, \dot{z}, \dot{\phi})^{\mathrm{T}} \tag{9.35}$$

Matrices $\mathbf{A}$ and $\mathbf{B}$ can be expressed as

$$\mathbf{A} = \begin{bmatrix} r\cos\phi - \rho_1 & 0 & 0 \\ 0 & r\cos\phi - \rho_2 & 0 \\ 0 & 0 & y + r - \rho_3 \end{bmatrix}, \tag{9.36}$$

$$\mathbf{B} = \begin{bmatrix} y & z + r\sin\phi & r(\rho_1\sin\phi + z\cos\phi) \\ y & z - r\sin\phi & r(\rho_2\sin\phi - z\cos\phi) \\ y + r - \rho_3 & z & 0 \end{bmatrix}. \tag{9.37}$$

The Jacobian matrix of the manipulator can be written as

$$\mathbf{J} = \mathbf{A}^{-1}\mathbf{B} \quad \text{or} \quad \mathbf{K} = \mathbf{J}^{-1} = \mathbf{B}^{-1}\mathbf{A}. \tag{9.38}$$

9.5.1.2 Optimization

The purpose of optimization design is to enhance the performance indices by adjusting the structure parameters. We propose the mean value and the standard deviation of the global stiffness as the design indices in this paper. It is noted that the trace of the matrix is an invariant of the matrix, so the distribution of the system stiffness (matrix) is the distribution of the trace. The mean value represents the average stiffness of the parallel robot manipulator over the workspace, while the standard deviation indicates the stiffness variation relative to the mean value. In general, the higher the mean value the less the deformation, and the lower the standard deviation the more uniform the stiffness distribution over the workspace. In this paper, the suitability of these design indices for the system stiffness will be examined by developing their relationship with the stiffness of links and joints. We will further study a design optimization based on the stiffness indices. A multiobjective optimization problem will be defined. Ideally, it may require that the mean value should be a maximum, but the standard deviation is a minimum. However, these two goals could be in conflict, so a trade-off process (e.g., Pareto set theory) will be considered.

The goal of structure parameters design, which is also called dimensional synthesis, is to confirm the best geometric configuration according to objective function and geometric restriction.

Since only a few geometric parameters can be handled due to the lack of convergence, this arises from the fact that traditional optimization methods use a local search by a convergent stepwise procedure, e.g., gradient, Hessians, linearity, and continuity, which compares the values of the next points and moves to the relative optimal points [60]. Global optima can be found only if the problem possesses certain convexity properties which essentially guarantee that any local optimum is a global optimum. In other words, conventional methods are based on a point-to-point rule; it has the danger of falling in local optima. The genetic algorithms are based on the population-to-population rule; it can escape from local optima.

For the implementation of genetic algorithms, one problem is how to model the objective function. It is very difficult and time-consuming exercise especially when the parameters are multifarious and the objective functions are too complex that genetic algorithm cannot work well based on the analytical expression of the performance indices. In artificial neural networks implementation, knowledge is represented as numeric weights, which are used to gather the relationships between data that are difficult to realize analytically. The network parameters can be iteratively adjusted to minimize the sum of the squared approximation errors using a gradient descent method, thereby being utilized to represent the system stiffness for the 3-dof parallel manipulator.

Stiffness is a very important factor in many applications including machine tool design, as it affects the precision of machining. Induced vibration is explicitly linked to machine tool stiffness. For a metal-cutting machine tool, high stiffness allows higher machining speeds and feeds while providing the desired precision, thus reduces vibration (such as chatter). Therefore, to build and study a general stiffness model of parallel mechanisms is very important for machine tool design.

From the viewpoint of mechanics, the stiffness is the measurement of the ability of a body or structure to resist deformation due to the action of external forces. The stiffness of a parallel mechanism at a given point of its workspace can be characterized by its stiffness matrix. This matrix relates the forces and torques applied at the gripper link in Cartesian space to the corresponding linear and angular Cartesian displacements.

The velocity relationship of parallel mechanisms can be written as

$$\dot{\boldsymbol{\theta}} = \mathbf{J}\dot{\mathbf{x}}, \tag{9.39}$$

where $\dot{\boldsymbol{\theta}}$ is the vector of joint rates and $\dot{\mathbf{x}}$ is the vector of Cartesian rates, a six-dimensional twist vector containing the velocity of a point on the platform and its angular velocity. Matrix $\mathbf{J}$ is the Jacobian matrix in (9.23).

From (9.24), one can conclude that

$$\boldsymbol{\delta}\theta = \mathbf{J}\delta\mathbf{x}, \tag{9.40}$$

where $\boldsymbol{\delta}\theta$ and $\boldsymbol{\delta}\mathbf{x}$ represent joint and Cartesian infinitesimal displacements, respectively. Then, one can get the stiffness of this mechanism using the principle of kinematic/static duality. The forces and moments applied at the gripper under static conditions are related to the forces or moments required at the actuators to maintain the equilibrium by the transpose of the Jacobian matrix $\mathbf{J}$. One can then write

$$\mathbf{F} = \mathbf{J}^{\mathrm{T}}\mathbf{f}, \tag{9.41}$$

where $\mathbf{f}$ is the vector of actuator forces or torques and $\mathbf{F}$ is the generalized vector of Cartesian forces and torques at the gripper link. The actuator forces and displacements can be related by Hooke's law, one has

$$\mathbf{f} = \mathbf{K}_{\mathrm{J}}\boldsymbol{\delta\theta}. \tag{9.42}$$

Here $\mathbf{K}_{\mathrm{J}}$ is the joint stiffness matrix of the parallel mechanism, with $\mathbf{K}_{\mathrm{J}} = \mathrm{diag}[k_1, \ldots, k_n]$, where each of the actuators in the parallel mechanism is modeled as an elastic component. k_i is a scalar representing the joint stiffness of each actuator, which is modeled as linear spring. Substituting (9.25) into (9.26), one obtains

$$\mathbf{f} = \mathbf{K}_{\mathrm{J}}\mathbf{J}\delta\mathbf{x}. \tag{9.43}$$

Then, substituting (9.28) into (9.26), yields

$$\mathbf{F} = \mathbf{J}^{\mathrm{T}}\mathbf{K}_{\mathrm{J}}\mathbf{J}\delta\mathbf{x}. \tag{9.44}$$

Hence, $\mathbf{K}_{\mathrm{C}}$, the stiffness matrix of the mechanism in the Cartesian space is then given by the following expression

$$\mathbf{K}_{\mathrm{C}} = \mathbf{J}^{\mathrm{T}}\mathbf{K}_{\mathrm{J}}\mathbf{J}. \tag{9.45}$$

Particularly, in the case for which all the actuators have the same stiffness, i.e., $k = k_1 = k_2 = \cdots = k_n$, then (9.30) will be reduced to

$$\mathbf{K}_C = k\mathbf{J}^T\mathbf{J}. \tag{9.46}$$

Furthermore, the diagonal elements of the stiffness matrix are used as the system stiffness value. These elements represent the pure stiffness in each direction, and they reflect the rigidity of machine tools more clearly and directly. The objective function for mean value and standard deviation of system stiffness can be written as:

$$\mu\text{-stiffness} = \mathbf{E}(\text{tr}(\mathbf{K}_C)), \tag{9.47}$$

$$\sigma\text{-stiffness} = \mathbf{D}(\text{tr}(\mathbf{K}_C)), \tag{9.48}$$

where $\mathbf{E}(\cdot)$ and $\mathbf{D}(\cdot)$ represent the mean value and the standard deviation, respectively, and tr is the trace of the stiffness matrix $\mathbf{K}_C$.

In order to obtain the optimal system stiffness of the 3-dof parallel manipulator, three geometrical parameters are selected as optimization parameters. The vector of optimization variables is

$$\mathbf{s} = \{L, h, r\}, \tag{9.49}$$

where L is link length, h is the height of the moving platform, r is the radii of fixed platform, and their bounds are

$$L \in [1, 2]\,\text{m}, \;\; h \in [0.5, 0.8]\,\text{m}, \;\; r \in [0.1, 0.3]\,\text{m} \tag{9.50}$$

The standard back propagation learning algorithm, as the most popular training method for feed-forward neural network, is based on the principle of steepest descent gradient approach to minimize a criterion function representing the instantaneous error between the actual outputs and the predicted outputs.

The criterion function can be expressed as follows:

$$E = \sum_{k=1}^{K} E_k^2 = \frac{1}{2NK}\sum_{k=1}^{K}\left[\sum_{i=1}^{N}(y_{ik} - t_{ik})^2\right], \tag{9.51}$$

where K is the number of output neurons, N is the vector dimension, and y_{ik}, t_{ik} are the predicted outputs and actual outputs of the kth output neuron of the ith input dimension, respectively.

The basic training steps of back propagation neural network are included as follows:

1. Initialize the weights and bias in each layer with small random values to make sure that the weighted inputs of network would not be saturated.
2. Confirm the set of input/output pairs and the network structure. Set some related parameters, i.e., the desired minimal , the maximal iterative times, and the learning speed.

3. Compare the actual output with desired network response and calculate the deviation.
4. Train the updated weights based on criterion function in each epoch.
5. Continue the above two steps until the network satisfy the training requirement.

Figure 9.24 shows the topology of network developed as the objective function to model the analytical solution of mean value of the system stiffness (μ-stiffness). In this case, two hidden layers with sigmoid transfer function are established in which eight neurons exist, respectively. The input vectors are the random arrangement of discretization values from the three structure variables.

Figure 9.25 illustrates the training result using standard back propagation learning algorithm, where the green curve denotes the quadratic sum of output errors with respect to ideal output values. After training for 474 times, target goal error is arrived.

The genetic algorithms can be implemented to search for the best solutions after the trained neural network is ready for the objective function. To avoid the time-consuming iterative operation using traditional technologies, the issue of stiffness optimization can be converted into network optimization. Figure 9.26 shows the evolution process of the best individual (network) based on genetic algorithms. The optimal μ-stiffness value is 2,218.

The evolution of μ-stiffness arises from the optimization of architecture and behavior variables in the implementation process of genetic algorithm as shown in Fig. 9.27. By adjusting the three parameters simultaneously with genetic operators such as selection, crossover, and mutation, the optimal objective is obtained. The final values of three parameters searched by genetic algorithm are

$$\mathbf{s} = \{L, h, r\} = \{1.5718\,\mathrm{m}, 0.78579\,\mathrm{m}, 0.18845\,\mathrm{m}\}. \tag{9.52}$$

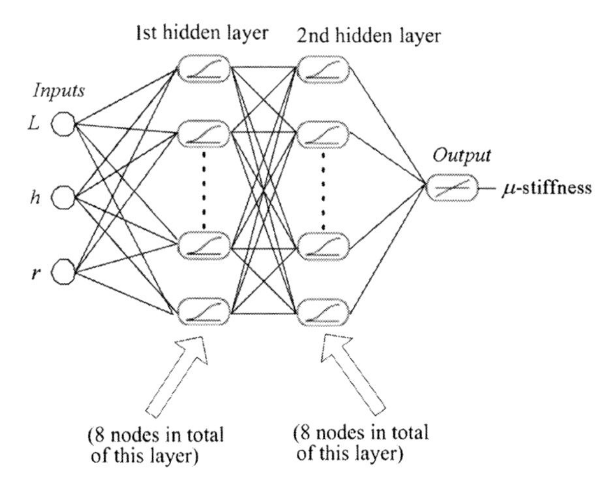

Fig. 9.24 The topology of feed forward neural network for the solution of μ-stiffness

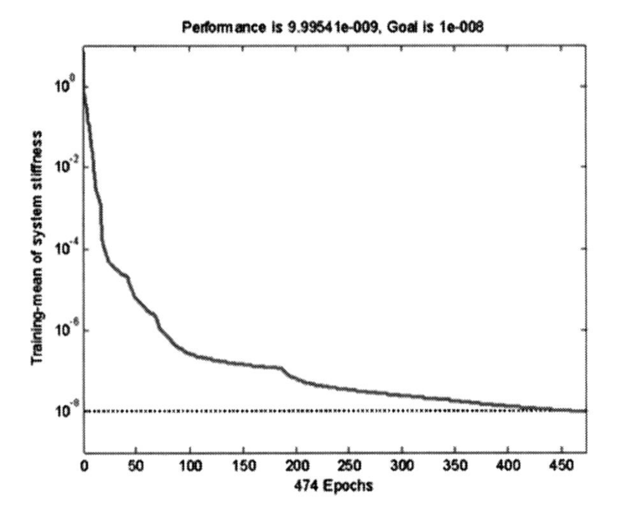

Fig. 9.25 Network training of the objective function about μ-stiffness

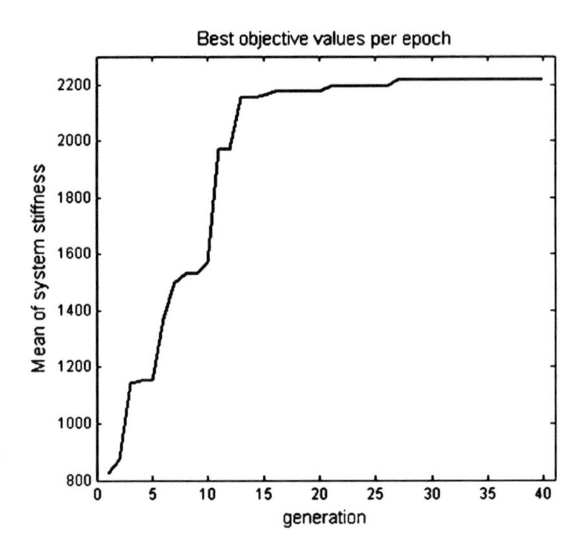

Fig. 9.26 The evolution of μ-stiffness

The topology of neural networks is similar with the above instance for the optimization of variance of system stiffness (σ-stiffness), in which the only difference is that there are 12 neurons exist in each hidden layer. Figure 9.28 illustrates the training result of the objective function about σ-stiffness with the back propagation algorithm, where the solid curve denotes the quadratic sum of output errors with respect to ideal output values. After training for 294 times, target goal error is arrived. The optimization process of σ-stiffness with genetic algorithm is illustrated in Fig. 9.29. After global stochastic search for 40 generations, optimal σ-stiffness value is convergent at 0.47. The evolution of geometrical parameters for σ-stiffness optimization is described in Fig. 9.30. Compared with Fig. 9.27, it can be found that

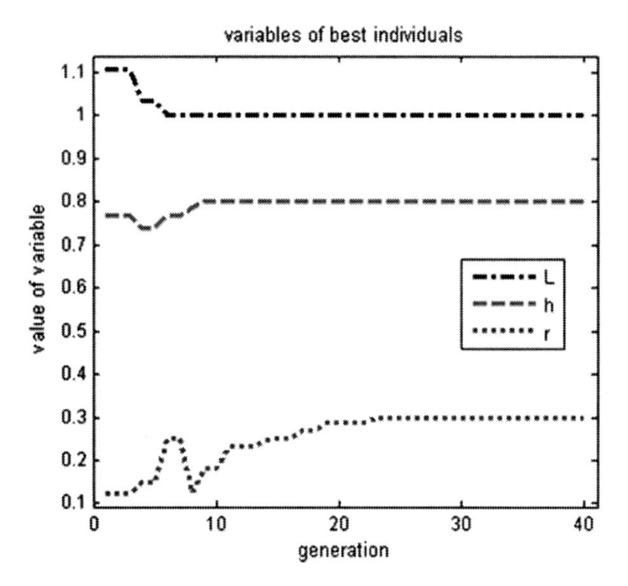

Fig. 9.27 The evolution of geometrical parameters for μ-stiffness optimization

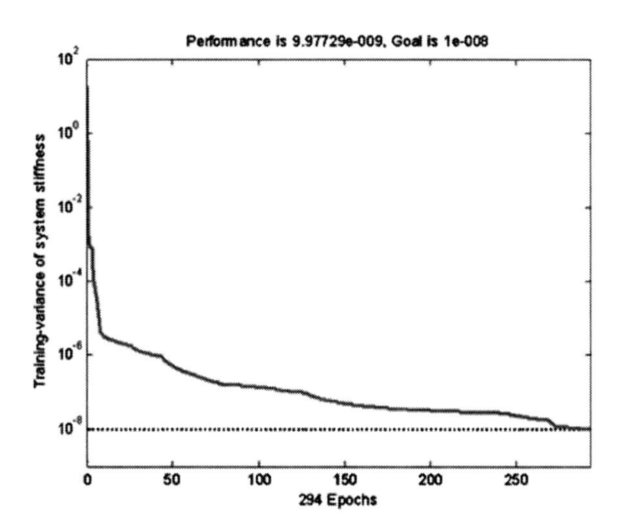

Fig. 9.28 Network training of the objective function about σ-stiffness

the corresponding convergent points of the three parameters in these two figures are not the same, i.e.,

$$\mathbf{s} = \{L, h, r\} = \{1.5718\,\mathrm{m}, 0.78579\,\mathrm{m}, 0.18845\,\mathrm{m}\}. \tag{9.53}$$

In other words, the two objective functions are conflicted with each other. This issue will be addressed in the following section where multiobjective optimization

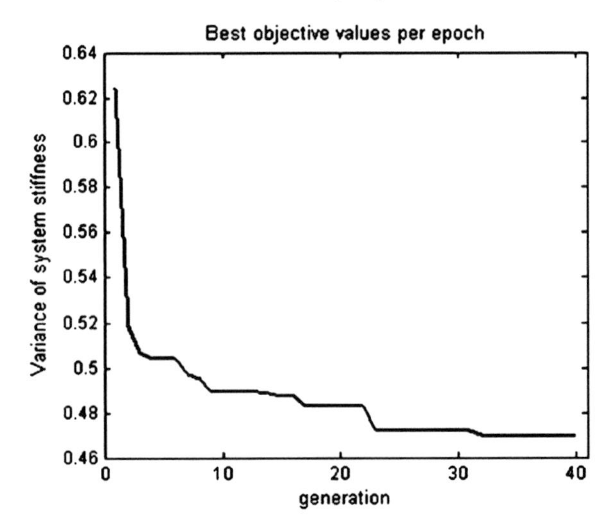

Fig. 9.29 The evolution of σ-stiffness

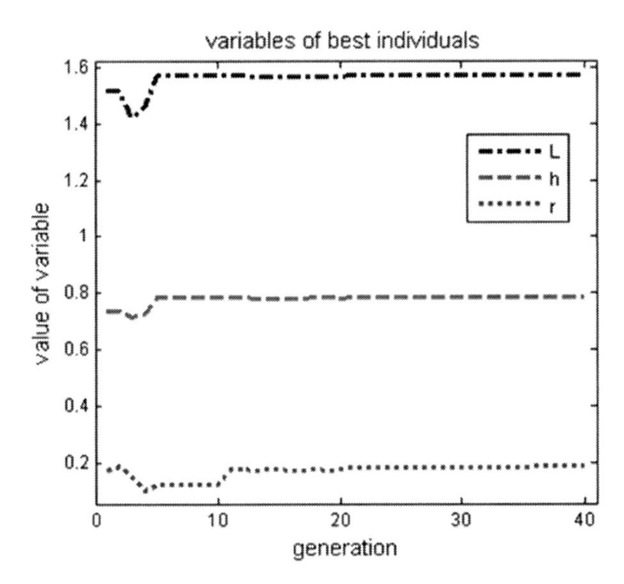

Fig. 9.30 The evolution of geometrical parameters for σ-stiffness optimization

based on Pareto-optimal solution is conducted. Multiobjective optimization problems consist of simultaneously optimizing several objective functions that are quite different from those of single-objective optimization. One single global optimal search is enough for single-objective optimization task. However, in a multiobjective optimization problem, it requires to find all possible tradeoffs among multiple objective functions that usually conflict with each other. The set of Pareto-optimal solutions is generally used for decision maker.

The basic concept of multiobjective optimization is the concept of domination [115]. In the issue of maximizing the k objective functions, decision vector (sets of variable) $x*$ is the Pareto-optimal solution if no other decision vectors satisfy both the following conditions:

$$f_i(x) \geq f_i(x^*), \ \forall i \in \{1, 2, \ldots, k\}, \tag{9.54}$$

$$f_j(x) > f_j(x^*), \ \exists j \in \{1, 2, \ldots, k\}. \tag{9.55}$$

With the same method, if both of the following conditions are true, decision vector x dominates y in the maximization issue, noted by $x > y$. That is:

$$f_i(x) \geq f_i(y), \ \forall i \in \{1, 2, \ldots, k\}, \tag{9.56}$$

$$f_j(x) > f_j(x^*), \ \exists j \in \{1, 2, \ldots, k\}. \tag{9.57}$$

According to the above formulae, Pareto-optimal set can be defined as: if there is no solution in the search space which dominates any member in the set P, then the solutions belonging to the set P constitute a global Pareto-optimal set. The Pareto-optimal set yields an infinite set of solutions, from which the desired solution can be chosen. In most cases, the Pareto-optimal set is on the boundary of the feasible region. Typical application of Pareto-based approach can be found in [64]. Since the implementation of genetic algorithms – including selection, crossover, and mutation operation – focuses on the whole colony which is consisted by all individuals, and generally, Pareto-optimal solutions for multiobjective optimization issues are a multidimensional set. Therefore, genetic algorithms are the effective method to address the Pareto-optimal solutions of multiobjective optimization issues.

Following initial parameters of Pareto-based genetic algorithms are set before implementation:

Number of subpopulation = 3
Number of individuals in each subpopulation = 50, 30, 40
Mutation range = 0.01
Mutation precision = 24
Max generations for algorithm termination = 200

After optimization, the possible optimal solutions in the whole solution space are obtained without combining all the objective functions into a single one by weighting factors. Figure 9.31 shows the Pareto-optimal frontier sets in which the designers can intuitively determine the final solutions depending on their preferences. Hence, the analysis process and cycle time is reduced. From this figure, the trade-off between the objectives of μ-stiffness and σ-stiffness is demonstrated in the distributing trend of these Pareto points for selecting compromisingly. If any other pair of design variables is chosen from the upper/right area of Figure 9.31, its corresponding values will locate an inferior point with respect to the Pareto frontier. Besides, the lower/left side is the inaccessible area of all the possible solution pairs. That is why Pareto solutions are called Pareto-optimal frontier sets. Figure 9.32 illustrates the

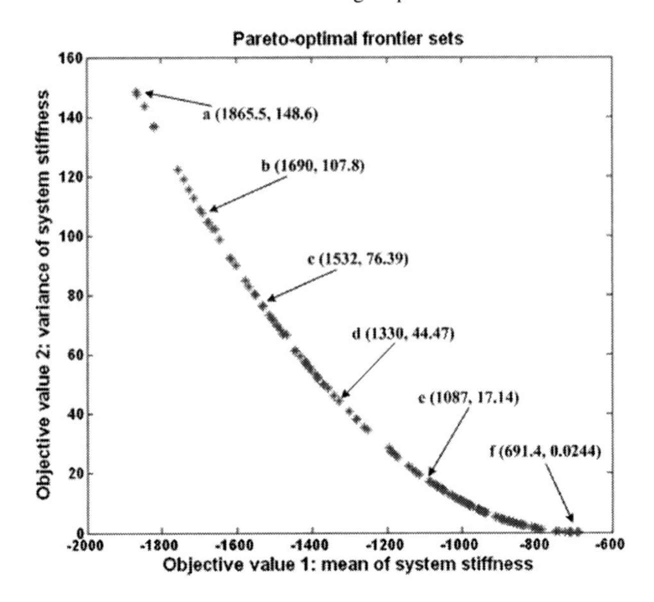

Fig. 9.31 Pareto-optimal frontier sets

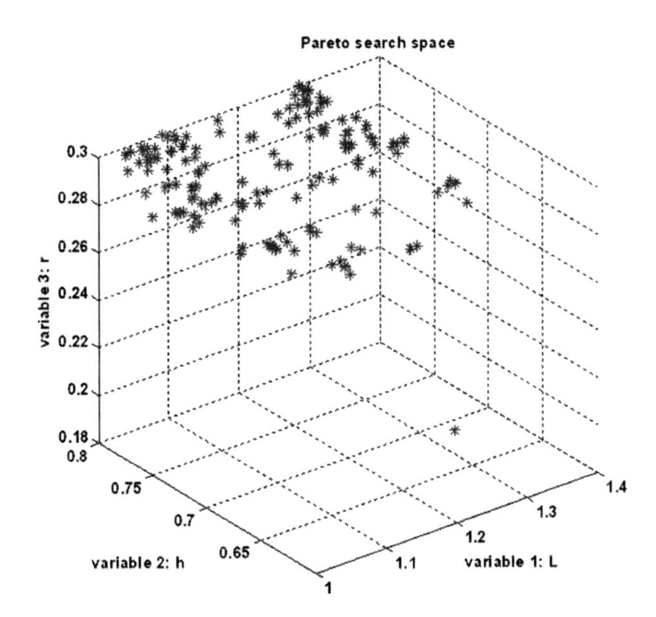

Fig. 9.32 Pareto search space: best individuals at end of optimization

solution distribution in the three-dimensional Pareto search space. It shows that a set of satisfied optimal solutions which provide enough information about alternative solutions for the decision maker with great diversity can be obtained with Pareto-based genetic algorithms. Therefore, the simulation shows the efficiency of the proposed single-objective and multiobjective optimization design methodology of the 3-dof parallel manipulator.

9.5.2 *Case Study 2: Tripod Compliant Parallel Micromanipulator*

9.5.2.1 Structure Description

As shown Fig. 9.33, the mechanism is treated as two separated components: a parallel mechanism with three 7-dof S$\underline{P}$S legs and a branched chain with a 3-dof RPR passive leg. Compared to common tripod parallel mechanism with no passive leg, more advantages can be found in this design. First, its motion comes solely from actuation of the prismatic components which facilitates in control and analysis of the motion paths of the mechanism. Second, it provides an adequate working envelope and due to the nature of the deformation incurred by the joints, it has a relatively high duration. Finally, it has a high accuracy but still leaves room for improvement due to the fact that input angles are used as the reliant factor. This also further complicates the motion control slightly and the Jacobian matrix. Since rotation about the z-axis is not required, this leg constrains that motion.

It also minimizes the torque and force on the other components of the mechanism. However, it must sustain induced bending and torsion created by external loads on the moving platform. The additional leg provides further stiffness at the end effecter and increases the overall precision and repeatability.

The repeated use of spherical joints has both advantages and disadvantages associated with it. It can cut down on the cost of the equipment because there are

Fig. 9.33 Conceptual design of the parallel compliant micro-manipulator

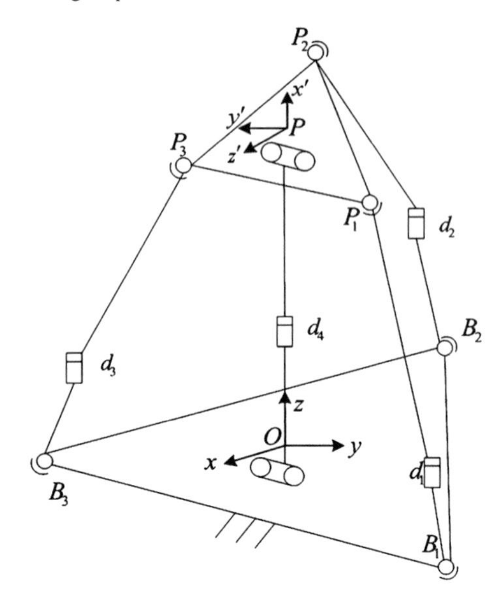

Fig. 9.34 Kinematical structure of a 3S<u>P</u>S+RPR parallel mechanism

less different types of parts required. This also facilitates the manufacturing and assembly processes. However, the use of spherical joints does limit the motion and working envelope of the mechanism. Because the motion required of the device is so tiny, this drawback can be neglected. In terms of analysis and motion control, it appears to be suitable also. Each of the two mechanisms is solved separately to determine the inverse kinematics.

There are some important assumptions that must be noted before progressing with the inverse kinematic modeling. It is assumed that each leg is driven by one actuator which drives the prismatic joint. It is supposed that the centers of the joints which form a triangle on both the base and moving platform are located on circles. The centers of these circles serve as the origins for both the fixed reference frame, denoted by $O\{x, y, z\}$ in Fig. 9.34 and a moving coordinate frame, denoted by $P\{x', y', z'\}$. The points of attachment of the revolute joints at the base are expressed by B_i and of the spherical joints at the moving platform by P_i where $i = 1$, 2, 3. Points B_1, B_2, and B_3 lie on the x–y plane. Similarly, points P_1, P_2, and P_3 lie on the y'–z' plane. Furthermore, each platform is supposed to be an equilateral triangle.

9.5.2.2 Performance Indices Optimization

The goal of structure parameters design, which is also called dimensional synthesis, is to confirm the best geometric configuration according to objective function and geometric restriction. To make sure that the parallel manipulator will possess well performance such as high system stiffness and dexterity, dimensional synthesis for optimization is one of the significant steps in the design process of parallel manipulators. Both the single-objective optimizing and multiobjective optimizing issues will be investigated in this section to demonstrate the validity of synthesis of radial basis function network (RBFN) and genetic algorithm for this case.

As one type of feed-forward neural networks that is different from common networks such as back propagation networks, RBFN has a special structure consisting of two layers: a nonlinear hidden layer and a linear output layer. Each of the units in hidden layer applies a fixed-feature detector which uses a specified kernel function (i.e., Gaussian, thin plate spline, or multiquadratic) to detect and respond to localized portions of the input vector space. The network output is a weighted linear summation of the output of the hidden neurons [52, 141]. One advantage of radial basis networks over BPNN is that the localized nature of the hidden layer response makes the networks less susceptible to weight loss. The RBFN is a universal function approximation approach that demonstrates more robustness and flexibility than traditional regression approaches such as polynomial fits. The RBFN works by choosing not just a single nonlinear function, but a weighted sum of a set of nonlinear functions (Fig. 9.35).

The kernel functions in the hidden layer produce a localized response to the input by using the distance between the input vector and the center associated with the hidden unit as the variable. Suppose the input sample $\mathbf{X} \in \mathbf{R}^n$, the corresponding output of RBFN is:

$$\phi_j(x) = k_j(\|\mathbf{X} - \mathbf{C}_j\|_2, \sigma), \tag{9.58}$$

where $\mathbf{C}_j$ is the center associated with the hidden node j and σ is the controlling coefficient of kernel function for hidden node j, which represent a measure of the spread of data. $\|\mathbf{X} - \mathbf{C}_j\|$ is a norm of $\mathbf{X} \to \mathbf{C}_j$ that is usually Euclidean, which denotes the distance between the input vector $\mathbf{X}$ and $\mathbf{C}_j$. k_j is a kernel function with radial symmetry, which achieves the unique maximum at the point of $\mathbf{C}_j$. Generally, Gaussian function is selected as the kernel function, namely,

$$k_j(\|\mathbf{X} - \mathbf{C}_j\|_2, \sigma) = \exp\left(-\frac{\|\mathbf{X} - \mathbf{C}_j\|_2^2}{2\sigma^2}\right). \tag{9.59}$$

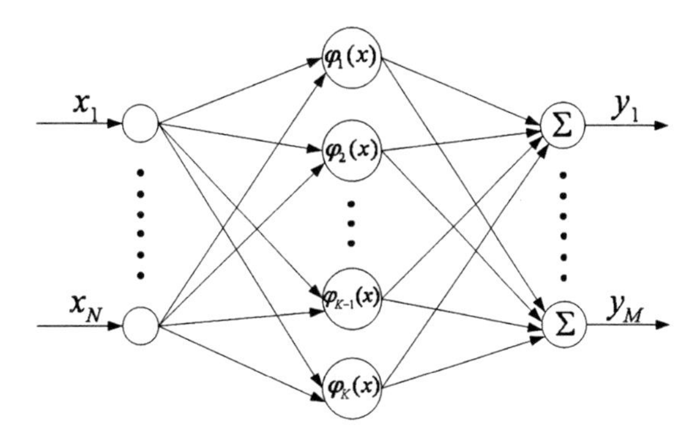

Fig. 9.35 The topology of RBFN

The response of each output node is calculated by a linear function of its input (including the bias), that is the output of hidden layer. Suppose that the number of hidden neurons and output neurons is K and M, respectively, the output value y_m of the mth output neuron for the input variable X can be represented by the following equation:

$$y_m = \sum_{i=1}^{K} w_{mi} k_j(\|\mathbf{X} - \mathbf{C}_j\|, \sigma), \tag{9.60}$$

where w_{mi}, which is adjusted to minimize the mean square error of the net output, is the weight between the mth output neuron and the ith hidden neuron.

Most of the training algorithms for RBFN have been divided into two stages. First, using unsupervised learning algorithm, the centers for hidden layer nodes can be determined. After the centers are fixed, the widths are determined in a way that reflects the distribution of the centers and input patterns. The pseudoinverse learning algorithm yields improved performance at a fraction of the computational and structural complexity of existing gradient descent algorithms for net weights training.

According to (9.60), the expression of error cost function $E(W)$ is as:

$$\mathbf{E(W)} = \frac{1}{2}\|\mathbf{T} - \mathbf{Y}\|_F^2, \tag{9.61}$$

where T is the net target output and $\|\cdot\|_F$ represents the F-norm of the given matrix.

$$\mathbf{E(W)} = \frac{1}{2}\|\mathbf{T} - \mathbf{HW}\|_F^2 = \frac{1}{2}\sum_{j=1}^{M}\sum_{i=1}^{L}\left(t_{ij} - \sum_{k=1}^{m} h_{ik}w_{kj}\right)^2, \tag{9.62}$$

where H denotes the output matrix of hidden layer. The partial derivative of $E(W)$ can be calculated as:

$$\frac{\partial \mathbf{E(W)}}{\partial \mathbf{W}} = \left(\frac{\partial \mathbf{E}}{\partial w_{uv}}\right)_{M \times M} \tag{9.63}$$

$$\frac{\partial \mathbf{J}}{\partial w_{uv}} = -\sum_{j=1}^{L} h_{ui}^{\mathrm{T}}\left(e_{iv} - \sum_{i=1}^{M} h_{ik}w_{kj}\right). \tag{9.64}$$

Thus following equation can be deduced:

$$\left(\frac{\partial \mathbf{E}}{\partial w_{uv}}\right)_{M \times M} = \mathbf{H}^{\mathrm{T}}(\mathbf{T} - \mathbf{HW}). \tag{9.65}$$

To achieve zero error of the net output, it has

$$\mathbf{H}^{\mathrm{T}}\mathbf{HW} = \mathbf{H}^{\mathrm{T}}\mathbf{E}. \tag{9.66}$$

Then the optimal solution of weights W^* can be obtained as

$$\mathbf{W}^* = (\mathbf{H}^{\mathrm{T}}\mathbf{H})^{-1}\mathbf{H}^{\mathrm{T}}\mathbf{E} = \mathbf{H}^{+}\mathbf{E}, \tag{9.67}$$

where H^+ is the Moore–Penrose pseudoinverse of hidden output H.

Since only a few geometric parameters can be handled due to the lack of convergence, this arises from the fact that traditional optimization methods use a local search by a convergent stepwise procedure, e.g., gradient, Hessians, linearity, and continuity, which compares the values of the next points and moves to the relative optimal points [60]. Global optima can be found only if the problem possesses certain convexity properties which essentially guarantee that any local optimum is a global optimum. In other words, conventional methods are based on a point-to-point rule; it has the danger of falling in local optima.

The genetic algorithms are based on the population-to-population rule; it can escape from local optima. Genetic algorithms have the advantages of robustness and good convergence properties, i.e.,

1. They require no knowledge or gradient information about the optimization problems; only the objective function and corresponding fitness levels influence the directions of search.
2. Discontinuities present on the optimization problems have little effect on the overall optimization performance.
3. They are generally more straightforward to introduce, since no restrictions for the definition of the objective function exist.
4. They use probabilistic transition rules, not deterministic ones.
5. They perform well for large-scale optimization problems.

Genetic algorithms have been shown to solve linear and nonlinear problems by exploring all regions of state space and exponentially exploiting promising areas through mutation, crossover, and selection operations applied to individuals in the population. Therefore, genetic algorithms are suitable for the optimization problems studied here.

Although a single-population genetic algorithm is powerful and performs well on a wide variety of problems. However, better results can be obtained by introducing multiple subpopulations. Figure 9.36 shows the optimization rationale of the extended multipopulation genetic algorithm adopted in this research.

Multiobjective optimization problems consist of simultaneously optimizing several objective functions that are quite different from those of single-objective optimization. One single global optimal search is enough for single-objective optimization task. However, in a multiobjective optimization problem, it is required to find all possible tradeoffs among multiple objective functions that are usually conflicting with each other. The set of Pareto-optimal solutions is generally used for decision maker.

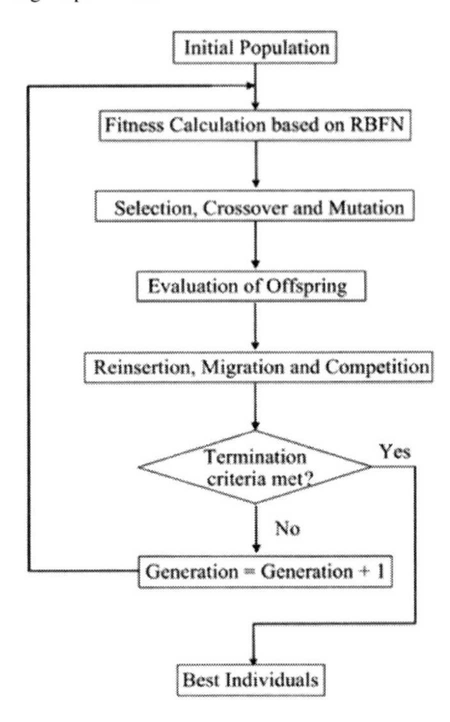

Fig. 9.36 Schematic representation of the optimization rationale based on genetic algorithms

Following initial parameters of Pareto-based genetic algorithms are set before implementation:

Number of subpopulation = 5
Number of individuals in each subpopulation = 110, 90, 90, 100, 110
Mutation range = 0.01
Mutation precision = 24
Max generations for algorithm termination = 80

Global stiffness (compliance), dexterity, and manipulability are considered together for the simultaneous optimization. After implementation, the possible optimal solutions in the whole solution space are obtained without combining all the objective functions into a single-objective function by weighting factors. Figure 9.37 shows the Pareto-optimal frontier sets in which the designers can intuitionistically determine the final solutions depending on their preferences. Hence, the analysis process and cycle time is reduced in large scale. From this picture, trade-off between the objectives of system stiffness, dexterity, and manipulability is demonstrated in the distributing trend of these Pareto points for selecting compromisingly. It shows that a set of satisfied optimal solutions which provide enough information about alternative solutions for the decision maker with great diversity can be obtained with Pareto-based genetic algorithms. Therefore, the simulation shows the efficiency of the proposed single-/multiobjective optimization methodology of the 3-dof parallel manipulator.

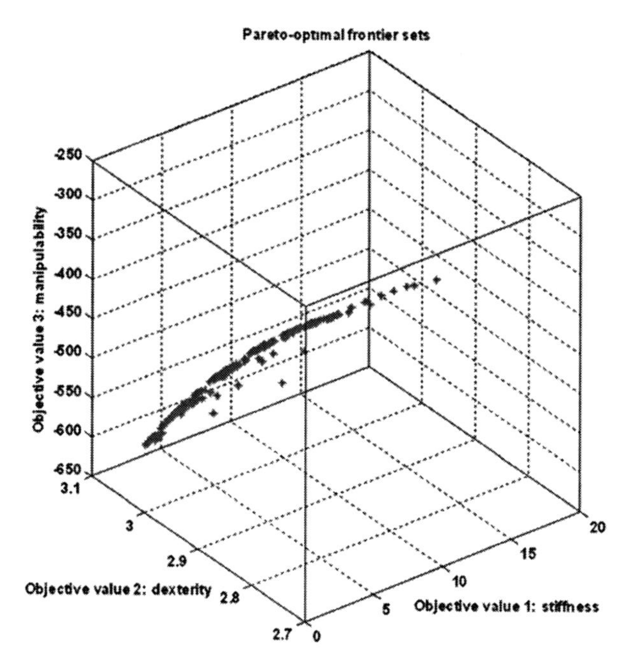

Fig. 9.37 Pareto-optimal solutions and Pareto frontier in the solution space

9.6 Conclusions

The kinetostatic model with its underlying design principles has been made more explicit through the implementation of optimization based on genetic algorithms in this chapter. A very remarkable implementation is the optimization of the Tricept machine tool family. After slightly adjusting the radius of the platform and the base, the total global stiffness can be improved 1.96 times. For the other mechanisms, the global stiffness are all obviously improved (normally 1.01–5.4 times). The kinetostatic model analyzed and obtained in previous chapters is employed for optimal structure design. From the results which have been achieved, it can be seen that the kinetostatic model can be applied for flexible mechanism analysis and global stiffness analysis and it can be further used as an optimization tool for parallel mechanisms. Moreover, the versatility of genetic algorithm compared to the conventional optimization methods is shown in this chapter; it is quite appropriate for dealing with multiparameters problem.

Chapter 10
Integrated Environment for Design and Analysis of Parallel Robotic Machine

10.1 Preamble

Because of the recent trend toward high-speed machining HSM, there is a demand to develop parallel kinematic machine with high dynamic performance, improved stiffness, and reduced moving mass [2, 11, 93, 148]. However, as researchers at Giddings and Lewis have indicated, full integration of standard automation components, CAD, and a user interface are required before making its parallel kinematic machine readily available for the general market. A virtual environment that can be used for PKM design, analysis, and simulation is urgently demanded. Several efforts have been done on this topic. Pritschow [122] proposed a systematic methodology for the design of different PKM topologies. Merlet [106] developed the software for the optimal design of a specific PKM class – Stewart platform-based mechanisms. Jin and Yang [79, 80] proposed a method for topology synthesis and analysis of parallel manipulators. Huang et al. [75] made some efforts on conceptual design of 3dof translational parallel mechanisms. Nevertheless, there is no complete virtual system existing for PKM design and analyze from the literature.

With the objective of developing a practical methodology and related virtual environment for PKM analysis and design, several activities have been conducted at Integrated Manufacturing Technologies Institute of National Research Council of Canada. PKM is a key component of reconfigurable manufacturing systems in different industrial sectors. It is very important for PKM designers to design and analysis the potential PKM with an integrated virtual environment before fabrication. The virtual environment is used for modeling, simulation, planning, and control of the proposed PKM.

An instance of a virtual parallel machine tool will include models of the machine tool and workpiece mechanics, the cutting process and the control system. The instance of a virtual parallel machine tool will be the reference model for an existing machine tool. A 3D virtual environment is both a visualization tool as well as an interface to the virtual machine tool or the actual machine tool. The software environment is also a design tool for constructing the modular components of a parallel machine tool as well as the integrated system design.

D. Zhang, *Parallel Robotic Machine Tools*, DOI 10.1007/978-1-4419-1117-9_10,

The objectives of the virtual environment are to develop:

1. A software environment for modeling, simulation, control of parallel machine tools
2. Machine tool simulations that can predict the geometry and surface finish of parts
3. Basic modeling and simulation capabilities of using the NRC 3-dof PKM as an example
4. Reconfigurable control systems
5. Systems for real-time inspection of machining operations and path planning in terms of singularity free and workspace verification.

10.2 Case Study

In general, a systematic design methodology for parallel kinematic machine design and analysis consists of two engines: a generator and an evaluator. Some of the functional requirements identified are transformed into structural characteristics. These structural characteristics are incorporated as rules in the generator. The generator defines all possible solutions via a combinational analysis. The remaining functional requirements are incorporated as evaluation criteria in the evaluator to screen out the infeasible solutions. This results in a set of candidate mechanisms. Finally the most promising candidate is chosen for product design. Therefore, the architecture of the virtual environment is illustrated in Fig. 10.1. It consists of several modules from conceptual design (selection of the most promising structure) to embodiment design, from kinematic/dynamic analysis (evaluation criteria) to design optimization, simulation and control. In the following, the key components of the system are described in detail.

For each of kinematic mechanisms, the kinematic chains involved may lead to several possibilities (serial, parallel, or hybrid). A preliminary evaluation of the mobility of a kinematic chain can be found from the Chebychev–Grbler–Kutzbach formula.

$$M = d(n - g - 1) + \sum_{i=1}^{g} f_i, \tag{10.1}$$

where M denotes the mobility or the system DOF, d is the order of the system ($d = 3$ for planar motion, and $d = 6$ for spatial motion), n is the number of the links including the frames, g is the number of joints, and f_i is the number of DOFs for the ith joint.

Kinetostatic analysis [170] is essential for PKMs that are used for metal cutting, which requires large forces. Higher stiffness, equivalently lower compliance, means little deformation, resulting in better surface finish and longer tool life. In this chapter, two global compliance indices are introduced, namely the mean value and the standard deviation of the trace of the generalized compliance matrix. The mean value represents the average compliance of the PKM over the workspace, while the standard deviation indicates the compliance fluctuation relative to the mean value.

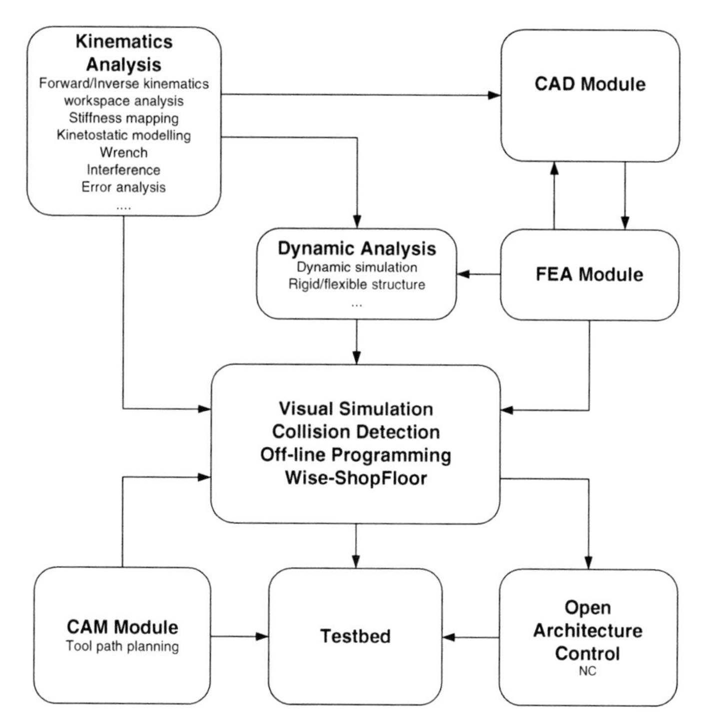

Fig. 10.1 Integrated environment for PKM design and analysis

In this model, it includes forward/inverse kinematics, workspace evaluation, velocity analysis, stiffness modeling, singularity analysis, and kinetostatic performance indices. The model will efficiently support the designer in the choice of a topological class of reconfigurable machine tools (RMT) and in the configuration of the machine belonging to that class.

An analysis package (MatLab) is used for comparative study of the characteristics of the RMT (i.e., manipulability, kinematics, stiffness mapping, workspace, kinetostatic, and dynamic analysis) to help the designer in selecting the most promising mechanism for a specific task. This tool will be used also for the development of control algorithms (i.e., inverse kinematic, interpolation, and real-time software collision checking). The kinetostatic model can be used to localize critical components, which mostly influence the global stiffness of the machine and further used for design optimization.

The dynamic model [174] will be used for accurate control and controller evaluation. The model should account for any factors that significantly affect the dynamic behavior of the parallel mechanisms. This includes joint friction, link flexibility, and eigenfrequencies used for the optimization of the model based servo (and for model verification). A general modeling method for the dynamic analysis of the PKM will be developed. In dynamic model, the mass/inertia, gravity of each component (including the links between the fixed base and the moving platform) will be included in this model. And this model will provide the relationship between the applied force/torque on the tool and the driving force/torque of the actuators as well as the constraint f orce/torque of all joints of the PKM.

Lagrange's formulation is used for dynamic modeling of the 3-dof NRC parallel kinematic machine. First, dynamic equations of the moving platform and the legs are formulated and then are assembled.

In the present work, there are many optimization parameters and complex matrix computations. Hence, it is very difficult to derive the analytical expressions for each stiffness element and workspace volume. Moreover, with traditional optimization methods, only a few geometric parameters can be handled due to the lack of convergence of the optimization algorithm when used with more complex problems [3]. This arises from the fact that traditional optimization methods use a local search by a convergent stepwise procedure (e.g., gradient, Hessians, linearity, and continuity), which compares the values of the next points and moves to the relative optimal points. Global optima can be found only if the problem possesses certain convexity properties that essentially guarantee that any local optimum is a global optimum. Classical optimization methods are based on point-to-point rules, and have the danger of falling in local optima, while the genetic algorithms are based on population-to-population rules, which allow them to escape from local optima. For this reason, genetic algorithms are selected as the best candidate for the optimization problems studied here.

Genetic algorithms have been shown to solve linear and nonlinear problems by exploring all regions of state space and exponentially exploiting promising areas through mutation, crossover, and selection operations applied to individuals in the population [4].

To use genetic algorithms properly, several parameter settings have to be determined, and they include chromosome representation, selection function, genetic operators, the creation of the population size, mutation rate, crossover rate, and the evaluation function. They are described in more detail as follows. Among these parameters, the chromosome representation is a basic issue for the GA representation, and it is used to describe each individual in the population of interest. For the reconfigurable parallel kinematic machine tools, the chromosomes consist of the architecture parameters (coordinates of the attachment points, coordinates of the moving platform, vertex distributions at base and moving platform, platform height, etc.) and behavior parameters (actuator stiffness, actuated link stiffness, etc.) of the mechanisms. The roulette wheel approach [4] is applied as a selection function.

A CAD module, used for the implementation from conceptual design phase to embodiment design phase – developed with Unigraphics, will support the designer for further finite element analysis and evaluation of the structure deformation and stress. The CAD module also can be used for simulation to check the interference. Meanwhile, it can be used to generate drawings for fabrication.

Reducing costs and increasing production throughout are two of the major challenges facing manufacturing companies today. Therefore, a CAM module is necessary for manufacturing systems. A CAM system is developed to meet the challenges with a set of capabilities for NC tool path creation, simulation, and verification. It delivers a single manufacturing solution capable of efficiently machining everything from holes to airfoils. The manufacturing application of Unigraphics allows one to interactively create NC machining programs, generate tool paths, visualize material removal, and post process. The CAM module can generate tool paths for several types of machining, such as planar and cavity milling, sequential milling, turning, surface contouring, drilling, thread milling, post builder, etc.

CAE module is used for finite element analysis (FEA). FEA software (Nastran) is required to do deformation and stress analysis of mechanical components included in the RMT. It can be further used to investigate mechanical properties and integrity of machined surfaces generated in high speed machining. It can be used to predict residual stresses and surface properties, determine the effects of diverse cutting-edge preparations and machining parameters (cutting speed, feed rate, and depth-of-cut) on the residual stress, distribution.

The Wise-ShopFloor is designed to provide users with a web-based and sensor-driven intuitive shop floor environment where real-time monitoring and control are undertaken. It utilizes the latest Java technologies, including Java 3D and Java Servlets, as enabling technologies for system implementation. Instead of camera images (usually large in data size), a physical device of interest (e.g., a milling machine or a robot) can be represented by a scene graph-based Java 3D model in an applet with behavioral control nodes embedded. Once downloaded from an application server, the Java 3D model is rendered by the local CPU and can work on behalf of its remote counterpart showing real behavior for visualization at a client side. It remains alive by connecting with the physical device through low-volume message passing (sensor data and user control commands). The 3D model provides users with increased flexibility for visualization from various perspectives, such as walk-through and fly-around that are not possible by using stationary optical cameras, whereas the largely reduced network traffic makes real-time monitoring, remote control, on-line inspection, and collaborative trouble-shooting practical for users on relatively slow hook-ups (e.g., modem and low-end wireless connections) through a shared Cyber Workspace [150].

By combining virtual reality models with real devices through synchronized real-time data communications, the Wise-ShopFloor allows engineers and shop floor managers to assure normal shop floor operations and enables web-based trouble-shooting – particularly useful when they are off-site.

The generalized stiffness matrix of a PKM relates a wrench including the forces and moments acting on the moving platform to its deformation. It represents how stiff the PKM is in order to withstand the applied forces and moments. By definition, the following relationship holds

$$\mathbf{w} = \mathbf{K}\delta\mathbf{x}, \tag{10.2}$$

where $\mathbf{w}$ is the vector representing the wrench acting on the moving platform, $\delta\mathbf{x}$ is the vector of the linear and angular deformation of the moving platform, and $\mathbf{K}$ is the generalized stiffness matrix. Vectors $\mathbf{w}$ and $\delta\mathbf{x}$ are expressed in the Cartesian coordinates $O - xyz$.

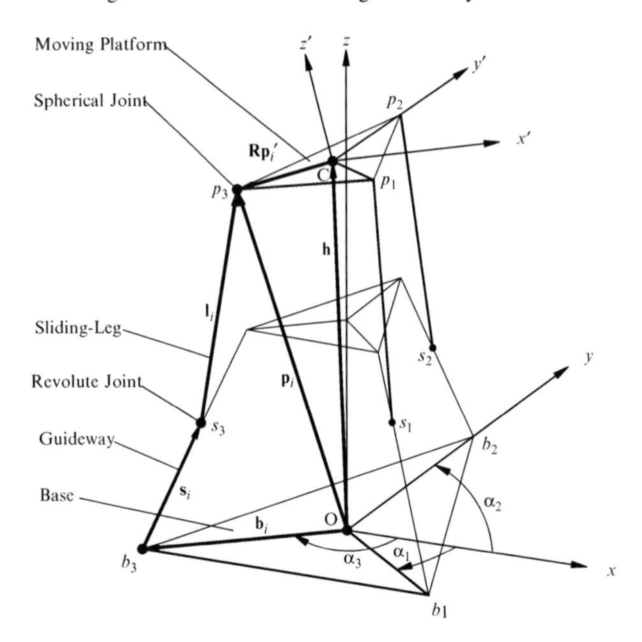

Fig. 10.2 Representation of NRC PKM

Since PKMs are parallel structures, the moving platform stiffness is a combination resulting from all serial chains including actuators. Figure 10.2 shows a schematic illustration of a 3 degree-of-freedom PKM with fixed-length legs that is built at NRC-IMTI. In this type of PKM, the moving platform is driven by sliding the fixed-length legs along the guideways. The advantages of the structure are the following: with this basic structure of parallel mechanism, it can be easily extended to 5-dof by adding two gantry type of guideways to realize the 5-dof machining; meanwhile, with the fixed length legs, one can freely choose the variety of leg forms and materials, and to use linear direct driver to improve the stiffness, and it is lack of heat sources to keep the precision in a high level, the stiffness is stable compare to variable legs.

Three types of compliance contribute to the deformation of the moving platform, namely actuator flexibility, leg bending, and axial deformation. A simple way of deriving the generalized stiffness matrix is to use the force relation and the infinitesimal motion relation as given below

$$\mathbf{w} = \mathbf{J}^{\mathrm{T}}\mathbf{f}, \tag{10.3}$$

$$\delta\mathbf{q} = \mathbf{J}\delta\mathbf{x}, \tag{10.4}$$

where $\mathbf{J}$ is the Jacobian matrix that relates the infinitesimal motion between the subserial chains and the moving platform, $\mathbf{f}$ is the vector representing forces in the sub-serial chains; $\delta\mathbf{q}$ is the vector representing infinitesimal motion of the subserial

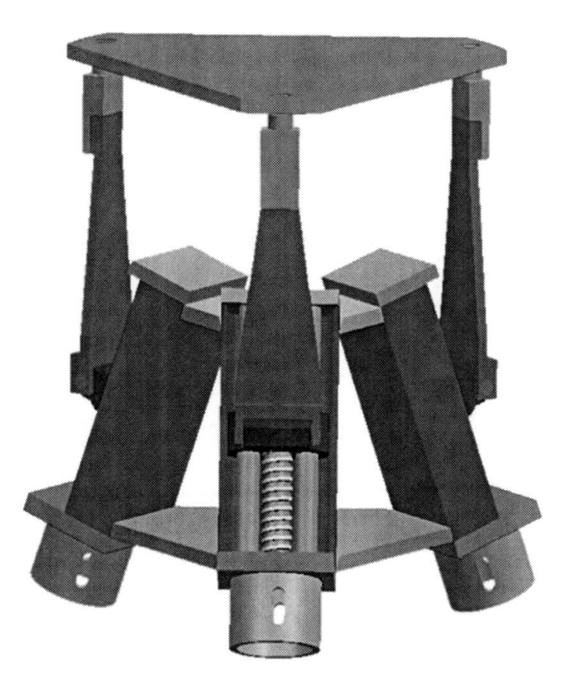

Fig. 10.3 CAD model of the tripod based PKM

chains. The infinitesimal motion of the subserial chains is referred to as the component deformation in the subserial chains. The component deformation would induce the forces, which are called the branch forces in the subserial chains (Fig. 10.3).

Considering the local stiffness in the subserial chains, denoted by $\overline{\mathbf{K}}$, the branch forces induced by the branch deformation can be written as

$$\mathbf{f} = \overline{\mathbf{K}}\delta\mathbf{q}. \tag{10.5}$$

The substitution of (10.4) and (10.5) into (10.3) yields

$$\mathbf{w} = \mathbf{K}\delta\mathbf{x}, \tag{10.6}$$

where the generalized stiffness matrix $\mathbf{K}$ is given as

$$\mathbf{K} = \mathbf{J}^{\mathrm{T}}\overline{\mathbf{K}}\mathbf{J}. \tag{10.7}$$

Equation (10.6) can be rewritten in terms of compliance as

$$\delta\mathbf{x} = \mathbf{C}\mathbf{w}, \tag{10.8}$$

where $\mathbf{C}$ is the generalized compliance matrix and $\mathbf{C} = \mathbf{K}^{-1}$. The generalized compliance matrix represents how much the moving platform would deform under the applied wrench $\mathbf{w}$.

When (10.8) is applied to consider the aforementioned three types of compliance, the following three types of the moving platform deformation would be induced

$$\delta \mathbf{x}_{\mathrm{t}} = \mathbf{C}_{\mathrm{t}} \mathbf{w}; \quad \delta \mathbf{x}_{\mathrm{b}} = \mathbf{C}_{\mathrm{b}} \mathbf{w}; \quad \delta \mathbf{x}_{\mathrm{a}} = \mathbf{C}_{\mathrm{a}} \mathbf{w}, \tag{10.9}$$

where subscripts t, b, and a indicate the deformation due to the torsion in the actuators, bending and axial deformation of the legs, respectively. Since these three deformations occur in a serial fashion, the total deformation can be considered as follows (based on the superposition theory in [142]).

$$\delta \mathbf{x} = \delta \mathbf{x}_{\mathrm{t}} + \delta \mathbf{x}_{\mathrm{b}} + \delta \mathbf{x}_{\mathrm{a}}. \tag{10.10}$$

This leads to the following compliance model,

$$\delta \mathbf{x} = \mathbf{C}_{\mathrm{G}} \mathbf{w}, \tag{10.11}$$

where the total generalized compliance matrix $\mathbf{C}_{\mathrm{G}}$ is given as

$$\mathbf{C}_{\mathrm{G}} = \mathbf{C}_{\mathrm{t}} + \mathbf{C}_{\mathrm{b}} + \mathbf{C}_{\mathrm{a}}. \tag{10.12}$$

In (10.12), $\mathbf{C}_{\mathrm{t}} = \mathbf{K}_{\mathrm{t}}^{-1}$, $\mathbf{C}_{\mathrm{b}} = \mathbf{K}_{\mathrm{b}}^{-1}$, and $\mathbf{C}_{\mathrm{a}} = \mathbf{K}_{\mathrm{a}}^{-1}$, and (10.12) can be rewritten as

$$\mathbf{C}_{\mathrm{G}} = \mathbf{K}_{\mathrm{t}}^{-1} + \mathbf{K}_{\mathrm{b}}^{-1} + \mathbf{K}_{\mathrm{a}}^{-1}, \tag{10.13}$$

where

$$\mathbf{K}_{\mathrm{t}} = \mathbf{J}_{\mathrm{t}}^{\mathrm{T}} \overline{\mathbf{K}}_{\mathrm{t}} \mathbf{J}_{\mathrm{t}}; \quad \mathbf{K}_{\mathrm{b}} = \mathbf{J}_{\mathrm{b}}^{\mathrm{T}} \overline{\mathbf{K}}_{\mathrm{b}} \mathbf{J}_{\mathrm{b}}; \quad \mathbf{K}_{\mathrm{a}} = \mathbf{J}_{\mathrm{a}}^{\mathrm{T}} \overline{\mathbf{K}}_{\mathrm{a}} \mathbf{J}_{\mathrm{a}}. \tag{10.14}$$

The total generalized stiffness matrix considering the three types of compliance can be written as

$$\mathbf{K}_{\mathrm{G}} = \mathbf{C}_{\mathrm{G}}^{-1}. \tag{10.15}$$

From (10.13), it can be seen that $\mathbf{C}_{\mathrm{G}}$ is defined by three different Jacobians and local stiffness corresponding to the three types of compliance.

The total generalized compliance matrix as defined in (10.12) does not have the appropriate units due to multiplication of the Jacobian. For this reason, a weighting matrix is applied to $\mathbf{C}_{\mathrm{G}}$ that becomes

$$\mathbf{C}_{\mathrm{W}} = \mathbf{W} \mathbf{C}_{\mathrm{G}} \mathbf{W}, \tag{10.16}$$

where the weighting matrix is defined as

$$\mathbf{W} = \mathrm{diag}(1, 1, 1, L, L, L). \tag{10.17}$$

In (10.16), L is a parameter with length unit. $\mathbf{C}_{\mathrm{W}}$ is a 6×6 matrix with the appropriate compliance units.

As shown in (10.12b), the compliance matrix is determined by the inverse of the stiffness matrix. Considering (10.13) and (10.15), the total generalized compliance matrix can be expressed as

$$\mathbf{C_W} = \mathbf{W}[(\mathbf{J}_t^T)\overline{\mathbf{K}}_t\mathbf{J}_t)^{-1} + (\mathbf{J}_b^T\overline{\mathbf{K}}_b\mathbf{J}_b)^{-1} + (\mathbf{J}_a^T\overline{\mathbf{K}}_a\mathbf{J}_a)^{-1}]\mathbf{W}. \tag{10.18}$$

For the prototype under study, it is an over-constrained kinematic system, and three Jacobains $\mathbf{J}_t$, $\mathbf{J}_b$ and $\mathbf{J}_a$ are 3×6 matrices. For this reason, the generalized inverse is applied and (10.18) is rewritten as

$$\mathbf{C_W} = \mathbf{C_{Wt}} + \mathbf{C_{Wb}} + \mathbf{C_{Wa}}, \tag{10.19}$$

where

$$\mathbf{C_{Wt}} = \mathbf{W}\mathbf{J}_t{}^+\overline{\mathbf{K}}_t{}^{-1}(\mathbf{J}_t{}^+)^T\mathbf{W}, \tag{10.20}$$

$$\mathbf{C_{Wb}} = \mathbf{W}\mathbf{J}_b{}^+\overline{\mathbf{K}}_b{}^{-1}(\mathbf{J}_b{}^+)^T\mathbf{W}, \tag{10.21}$$

$$\mathbf{C_{Wa}} = \mathbf{W}\mathbf{J}_a{}^+\overline{\mathbf{K}}_a{}^{-1}(\mathbf{J}_a{}^+)^T\mathbf{W}. \tag{10.22}$$

In (10.18), the superscript "+" indicates the generalized inverse matrix.

The generalized compliance matrix $\mathbf{C_G}$ varies over the PKM workspace. Conventional kinetostatic analysis methods, such as stiffness mapping, would require a large number of graphs to provide an overview of the stiffness variation. An alternative, however, could be based on statistical analysis. This method was proposed to evaluate the generalized mass matrix of PKMs over the workspace. On the basis of this concept, the mean value and the standard deviation of a selected parameter can be used to evaluate the variation over the workspace. Since the trace of the generalized compliance matrix is invariant, it is selected as a parameter for global kinetostatic analysis. The mean value and the standard deviation are defined as

$$\mu = E(\mathrm{tr}(\mathbf{C_W})), \tag{10.23}$$

$$\sigma = SD(\mathrm{tr}(\mathbf{C_W})), \tag{10.24}$$

where $E(\cdot)$ and $SD(\cdot)$ are the mean value and the standard deviation, and tr represents trace operation. The mean value represents the average compliance of the PKM over the workspace, while the standard deviation indicates the compliance fluctuation relative to the mean value. In general, the lower the mean value the lesser the deformation, and the lower the standard deviation the more uniform the compliance distribution over the workspace.

The method presented is generic and can be readily expanded to any kind of PKM or completely new topology of PKM. It is quite efficient for the conceptual design stage to rapidly configure and evaluate several configurations. It can be further used for geometry optimization.

The two global compliance indices introduced in (10.19) and (10.20) are used. In terms of $\mathbf{C}_{\mathrm{W}}$, they are rewritten as

$$\mu_{\mathrm{W}} = E(\mathrm{tr}(\mathbf{C}_{\mathrm{W}})), \tag{10.25}$$

$$\sigma_{\mathrm{W}} = SD(\mathrm{tr}(\mathbf{C}_{\mathrm{W}})). \tag{10.26}$$

With the two indices, analysis can be conducted to consider the effect of change in leg and actuator flexibility. To do so, the following two stiffness ratios are defined

$$\alpha_1 = k_b/k_t; \quad \alpha_2 = k_a/k_t, \tag{10.27}$$

where k_t is the actuator's stiffness and it is fixed, k_b is the stiffness induced by compliant link bending, k_a is the stiffness induced by axial deformation, 1 and 2 change from 0.5 to 2.5. If the two ratios are less than 1, it indicates that the actuator is relatively stiffer than the leg. If they are equal to 1, the leg and the actuator are equally stiff. If the ratios are larger than 1, then the leg is relatively stiffer than the actuator. The term α_1 is for the leg bending and the term α_2 for the leg axial deformation. In terms of α_1 and α_2, $\mathbf{C}_{\mathrm{Wa}}$ and $\mathbf{C}_{\mathrm{Wb}}$ can be rewritten as

$$\mathbf{C}_{\mathrm{Wb}} = (1/\alpha_1)\mathbf{W}\mathbf{J}_{\mathrm{b}}{}^{+}\overline{\mathbf{K}}_{\mathrm{t}}{}^{-1}(\mathbf{J}_{\mathrm{b}}{}^{+})^{\mathrm{T}}\mathbf{W}, \tag{10.28}$$

$$\mathbf{C}_{\mathrm{Wa}} = (1/\alpha_2)\mathbf{W}\mathbf{J}_{\mathrm{a}}{}^{+}\overline{\mathbf{K}}_{\mathrm{t}}{}^{-1}(\mathbf{J}_{\mathrm{a}}{}^{+})^{\mathrm{T}}\mathbf{W}. \tag{10.29}$$

To investigate the effect of change in leg and actuator flexibility on the global kinetostatic behavior of the prototype, the differences of the mean value and standard deviation are used and they are defined for the three types of compliance under consideration as

$$\Delta\mu_{\mathrm{Wt}} = E(\mathrm{tr}(\mathbf{C}_{\mathrm{W}} - \mathbf{C}_{\mathrm{Wt}})), \tag{10.30}$$

$$\Delta\mu_{\mathrm{Wb}} = E(\mathrm{tr}(\mathbf{C}_{\mathrm{W}} - \mathbf{C}_{\mathrm{Wb}})), \tag{10.31}$$

$$\Delta\mu_{\mathrm{Wa}} = E(\mathrm{tr}(\mathbf{C}_{\mathrm{W}} - \mathbf{C}_{\mathrm{W}a})), \tag{10.32}$$

$$\Delta\sigma_{\mathrm{Wt}} = SD(\mathrm{tr}(\mathbf{C}_{\mathrm{W}} - \mathbf{C}_{\mathrm{Wt}})), \tag{10.33}$$

$$\Delta\sigma_{\mathrm{Wb}} = SD(\mathrm{tr}(\mathbf{C}_{\mathrm{W}} - \mathbf{C}_{\mathrm{Wb}})), \tag{10.34}$$

$$\Delta\sigma_{\mathrm{Wa}} = SD(\mathrm{tr}(\mathbf{C}_{\mathrm{W}} - \mathbf{C}_{\mathrm{Wa}})). \tag{10.35}$$

The differences defined in (10.25)–(10.26) indicate the proximity of $\mathbf{C}_{\mathrm{Wt}}$, $\mathbf{C}_{\mathrm{Wb}}$ and $\mathbf{C}_{\mathrm{Wa}}$ to $\mathbf{C}_{\mathrm{W}}$. A smaller value would mean a larger contribution to the total generalized compliance.

Figure 10.4 shows the simulation result considering the full motion range of the moving platform in the vertical direction. For the purpose of examining the two ratios, three regions are divided. Region 1 is for $\alpha_1, \alpha_2 < 1$, corresponding to the case that the leg is more flexible than the actuator. Region 2 is for $\alpha_1, \alpha_2 > 1$, and $\alpha_2 > \alpha_1$, corresponding to the case that the actuator is more flexible than the leg, while for the leg, the bending is larger than the axial deformation. Region 3 is for

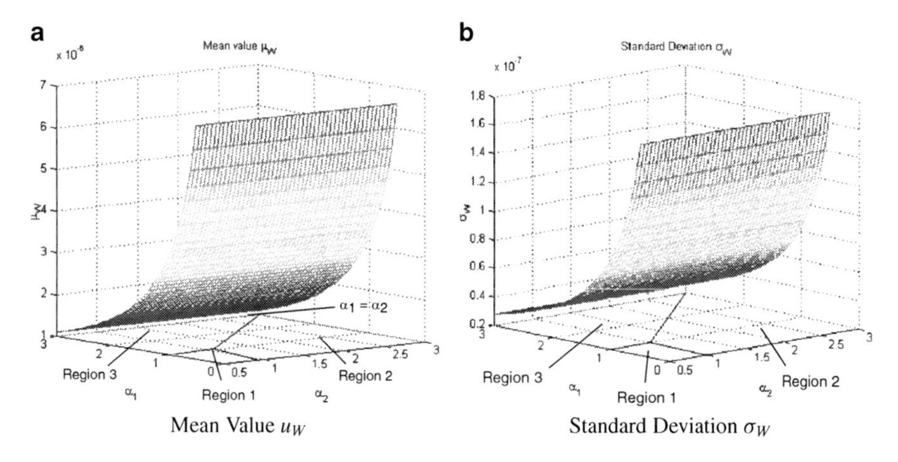

Mean Value u_W Standard Deviation σ_W

Fig. 10.4 Simulation results of global compliance mean value u_{W} and global standard deviation σ_{W}: Region 1: $\alpha_1, \alpha_2 < 1$, the leg is more flexible than the actuator. Region 2: $\alpha_1, \alpha_2 > 1$, and $\alpha_2 > \alpha_1$, the actuator is more flexible than the leg, while for the leg, the bending is larger than the axial deformation. Region 3: $\alpha_1, \alpha_2 > 1$, and $\alpha_1 > \alpha_2$, the actuator is more flexible than the leg, while for the leg, the axial deformation is larger than the bending

$\alpha_1, \alpha_2 > 1$ and $\alpha_1 > \alpha_2$, corresponding to the case that the actuator is more flexible than the leg, while for the leg, the axial deformation is larger than the bending. As shown in Fig. 10.4, for small α_1, i.e., when the bending is larger, it induces very large compliance at the moving platform. Hence the bending may be considered as a main factor.

The aim of optimization for PKMs is to minimize the global compliance and maximize the workspace volume. Therefore, it is a multiobjective optimization problem. The objective function is given as

$$\mathrm{val} = \max(1/\mu + 1/\sigma + V), \tag{10.36}$$

where μ represents the mean value of the trace of the global compliance matrix of the PKMs; σ is its standard deviation; and V is the workspace volume of the PKMs.

The methods for determination of the workspace can be found in the literature, and the method used here is the inverse kinematics-based method [103].

In the cases being studied, there are many parameters and the complicated matrix computation making it difficult to write out the analytical expressions for each stiffness element. Using traditional optimization methods, only a few geometric parameters [3] can be handled because of the lack of convergence of the optimization algorithm when used with more complex problems. This arises from the fact that traditional optimization methods use a local search by a convergent stepwise procedure (e.g., gradient, Hessians, linearity, and continuity), which compares the values of the next points and moves to the relative optimal points. Global optima can be found only if the problem possesses certain convexity properties that essentially guarantee that any local optima is a global optimum. In other words, conventional methods are based on point-to-point rule; it has the danger of falling in local optima.

Genetic algorithms (GAs) are powerful and broadly applicable stochastic search and optimization techniques based on evolutionary principles [5]. The genetic algorithms are based on population-to-population rule; it can escape from local optima. Therefore, genetic algorithms are the suitable for such optimization problems.

To use genetic algorithms properly, several parameter settings have to be determined, they are: chromosome representation, selection function, genetic operators, the creation of the population size, mutation rate, crossover rate, and the evaluation function. They are described in more detail as follows:

Chromosome representation. This is a basic issue for the GA representation; it is used to describe each individual in the population of interest. For the problem studied here, the chromosomes consist of the architecture parameters (coordinates of the attachment points, coordinates of the moving platform, vertex distributions at base and moving platform, platform height, etc.) and behavior parameters (actuator stiffness, actuated link stiffness, etc.) of the mechanisms.

Selection function. This step is a key procedure to produce the successive generations. It determines which of the individuals will survive and continue on to the next generation. In the study, the roulette wheel approach is applied.

Genetic operators. The operators are used to create new children based on the current generation in the population. Basically, there are two types of operators: crossover and mutation. Crossover takes two individuals and produces two new individuals, while mutation alters one individual to produce a single new solution.

Population size. The population size represents the number of individuals or chromosomes in the population.

Mutation rate. The mutation rate is defined as the percentage of the total number of genes in the population; it determines the probability that a mutation will occur. The best mutation rate is application dependent but for most applications is between 0.001 and 0.1. In the case studied, mutation rate is 0.1.

Crossover rate. The best crossover rate is application dependent but for most applications it is between 0.80 and 0.95. For the case studied, crossover rate is 0.85.

Evaluation functions. Evaluation functions are subject to the minimal requirement that the function can map the population into a partially ordered set.

Simulations are carried out on the 3-dof PKM prototype built at the Integrated Manufacturing Technologies Institute of the National Research Council of Canada as shown in Fig. 10.2. The base platform is a triangular plate with a side length of 245.5 mm and the moving platform is another triangular plate with a side length of 139.7 mm. The guideway length is 95.25 mm and the sliding leg length is 215.9 mm. The guideway angle relative to the vertical direction is 20°. The three stiffness values of the prototype are $k_t = 1.26e^{10}$N/m, $k_b = 3.13e^{10}$N/m, $k_a = 1.95e^{7}$N/m, and they are the same for the three subserial chains.

For the problem studied here, the chromosomes consist of the architecture parameters including coordinates of the attachment points, coordinates of the moving platform, link length, vertex distributions at base and moving platform, platform height, etc. Hence, the parameters selected for optimization are the following: $R_\mathrm{p}, R_\mathrm{m}, h_\mathrm{m}, \gamma$, where R_p is the radius of the moving platform; R_m is the radius of the middle plate; h_m is the height of the middle plate with respect to the base plate; γ is the rotation angle of the middle plate with respect to Cartesian Z-axis. And their bounds are

$$R_\mathrm{p} \in [60.96, 128.9]\ \mathrm{mm}, \quad R_\mathrm{m} \in [128.9, 304.8]\ \mathrm{mm},$$
$$h_\mathrm{m} \in [243.84, 365.76]\ \mathrm{mm}, \quad \gamma \in [-\pi/3, 0]\ \mathrm{rad},$$

Some other parameters are set as

$$P = 40,$$
$$G_{\max} = 100,$$

where P is the population and $G_{\max}$ the maximum number of generations.

One can rewrite the objective function (27) as

$$\mathrm{Val}(i) = W_\mu/\mu + W_\sigma/\sigma + W_r R_{\max} + W_z Z_{\mathrm{range}} \tag{10.37}$$

with $i = 1, 2, 3 \ldots 40$; $R_{\max}$ is the maximum radius of the workspace; Z_{range} is the range of movement in Cartesian Z-axis direction; and W_x is the weight factor for each entry. In this case, $W_\mu = 1$, $W_\sigma = 1$, $W_r = 0.1$; $W_z = 0.05$.

The objective functions are established and maximized to find the suitable geometric parameters (coordinates of the attachment points, coordinates of the moving platform, link length, vertex distributions at base and moving platform, platform height, etc.) and behavior parameters (actuator stiffness, actuated link stiffness, kinetostatic model stiffness, etc.) of the mechanisms. Since the objective function is closely related to the topology and geometry of the structure, and it is used to increase working volume to a certain value and to minimize the mean value and standard deviation of the global compliance matrix.

Once the objective function is written, a search domain for each optimization variable (lengths, angles, etc.) should be specified to create an initial population. The limits of the search domain are set by a specified maximum number of generations, since the GAs will force much of the entire population to converge to a single solution.

It is very difficult to optimize both global stiffness and workspace to their maximum values simultaneously, as larger workspace always leads smaller stiffness, and vice versa [159]. However, one can solve the problem by determining which item between workspace and stiffness is the dominant one for design and application, and maximize the dominant one while set the other one as a constant (but set as larger than the original). In this research, we set the workspace to a certain

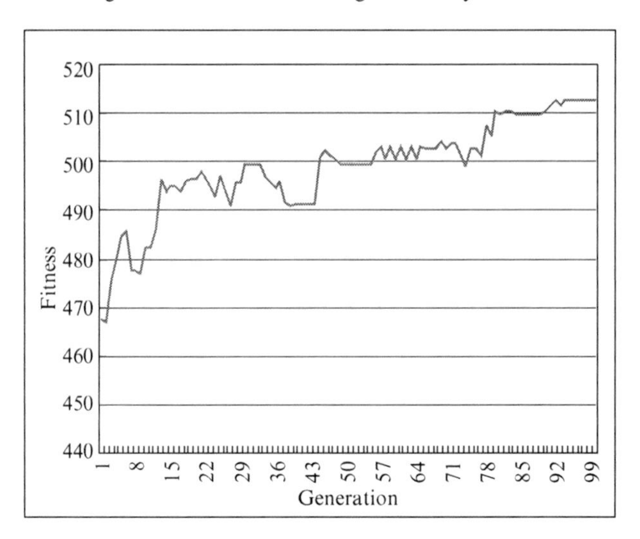

Fig. 10.5 The evolution of the performance of the NRC PKM

value, i.e., the radius of workspace is 304.8 mm, and then maximize the global stiffness. The algorithm converged at the 95th generation (Fig. 10.5). The optimized structure parameters are: $[R_p, R_m, h_m, \gamma] = [151, 259.8, 280.5, -0.1762]$, and the increased radius of the workspace is 304.8 mm; the range of movement along Z-axis is 304.8 mm; the sum compliance of the structure is 0.1568 mm/N.

The proposed methodology is implemented to design and optimization of the reconfigurable PKMs built at NRC-IMTI. The chromosomes consist of the architecture parameters including coordinates of the attachments at base and moving platforms, link length, platform height, coordinates of the moving platform. After optimization, the global stiffness is improved by a factor of 1.5, and workspace is increased 12%. A detailed example of industrial application is presented and analyzed in [170].

Java 3D is designed to be a mid to high-level fourth-generation 3D API [17]. What sets a fourth-generation API apart from its predecessors is the use of scene-graph architecture for organizing graphical objects in the virtual 3D world. Unlike the display lists used by the third-generation APIs (such as VRML, OpenInventor, and OpenGL), scene graphs can isolate rending details from users while offering opportunities for more flexible and efficient rendering. Enabled by the scene-graph architecture, Java 3D provides an abstract, interactive imaging model for behavior and control of 3D objects. Because Java 3D is part of the Java pantheon, it assures users ready access to a wide array of applications and network support functionality [136]. Java 3D differs from other scene graph-based systems in that scene graphs may not contain cycles. Thus, a Java 3D scene graph is a directed acyclic graph. The individual connections between Java 3D nodes are always a direct relationship: parent to child. Figure 10.6 illustrates a scene graph architecture of Java 3D for the NRC PKM. This test bed is a gantry system, which consists of an x-table and a 3-dof

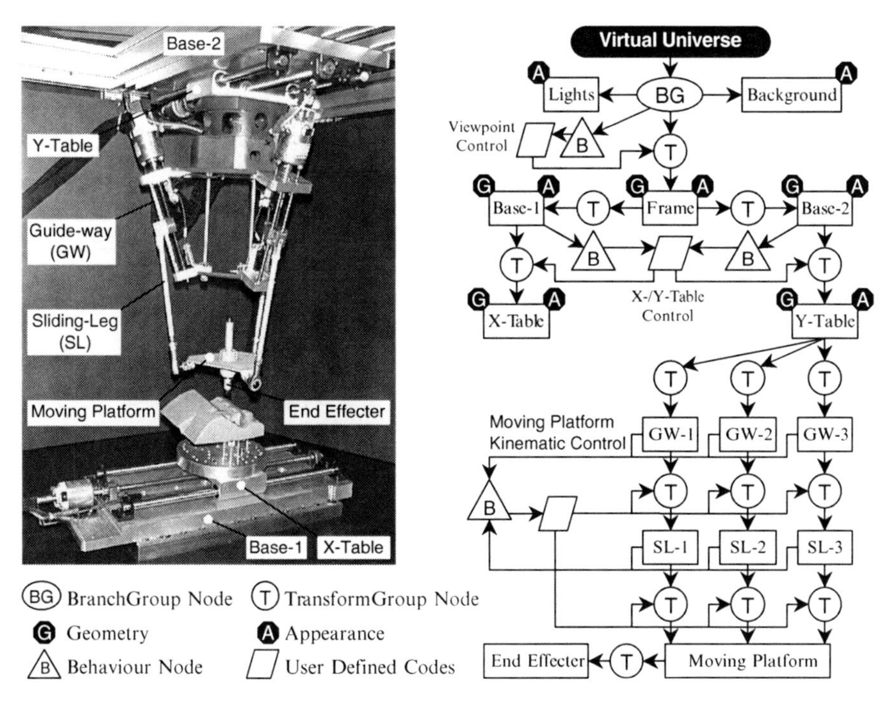

Fig. 10.6 Java 3D scene graph architecture for NRC PKM

PKM unit mounted on a y-table. The end effecter on the moving platform is driven by three sliding-legs that can move along three guide-ways, respectively.

As shown in Fig. 10.6, the scene graph contains a complete description of the entire scene with a virtual universe as its root. This includes the geometry data, the attribute information, and the viewing information needed to render the scene from a particular point of view. All Java 3D scene graphs must connect to a *Virtual Universe* object to be displayed. The *Virtual Universe* object provides grounding for the entire scene. A scene graph itself, however, starts with *BranchGroup* (BG) nodes. A *BranchGroup* node serves as the root of a branch graph of the scene graph. The *TransformGroup* nodes inside of a branch graph specify the position, the orientation, and the scale of the geometric objects in the virtual universe. Each geometric object consists of a *Geometry* object, an *Appearance* object, or both. The *Geometry* object describes the geometric shape of a 3D object. The *Appearance* object describes the appearance of the geometry (color, texture, material reflection characteristics, etc.). The behavior of the 3-dof PKM model is controlled by *Behavior* nodes, which contain user-defined control codes and state variables. Sensor data processing can be embedded into the codes for remote monitoring. Once applied to a *TransformGroup* node, the so-defined behavior control affects all the descending nodes. In this example, the movable objects (X-Table, Y-Table, and Moving Platform) are controlled by using three control nodes, for on-line monitoring/control and off-line simulation. As the Java 3D model is connected with its physical counterpart through the

a

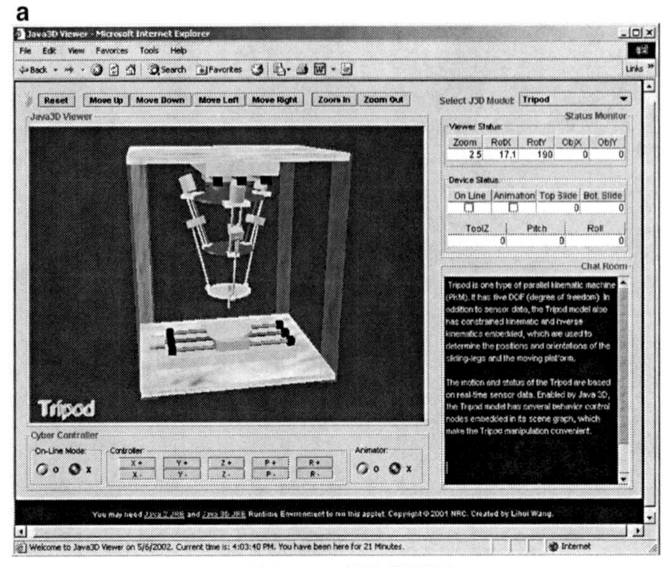

Initial state of NRC PKM

b

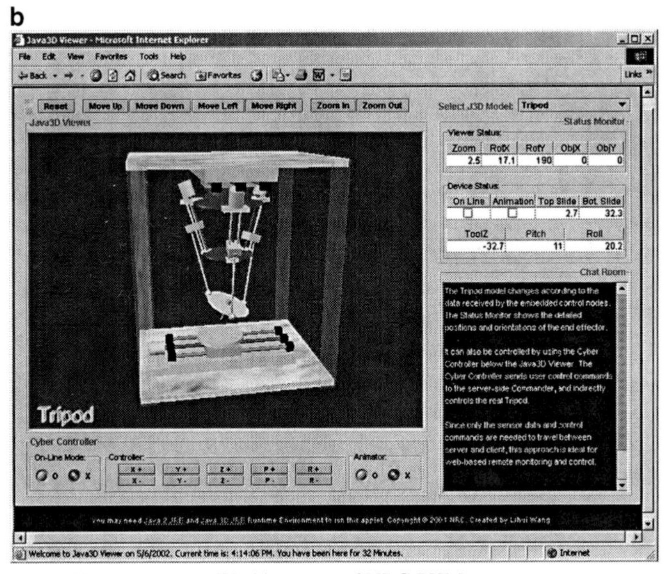

Working state of NRC PKM

Fig. 10.7 Web-based remote monitoring and control

control nodes by low-volume message passing (real-time sensor signals and control commands, etc.), it becomes possible to remotely manipulate the real NRC PKM through its Java 3D model.

Web-based remote device monitoring and control are conducted by using the *StatusMonitor* and *CyberController*, which communicate indirectly with the device controller through an application server. In the case of PKM monitoring and control, they are further facilitated by the kinematic models, to reduce the amount of data traveling between web browsers and the PKM controller. The required position z_c and orientations θ_x, θ_y of the moving platform are converted into the joint coordinates $s_i (i = 1, 2, 3)$ by the inverse kinematics for both Java 3D model rendering at client-side and device control at server-side. The three sliding-legs of the PKM are driven by three 24V DC servomotors combined with three lead screws. Each actuator has a digital encoder ($1.25\mu\text{m/count}$) for position feedback. The position data $s_i (i = 1, 2, 3)$ of the sliding-legs are multicast to the registered clients for remote monitoring, while only one user at one time is authorized to conduct remote control. A sampling rate of 1 kHz is used for the case study. Figure 10.7 shows how the PKM is manipulated from one state to another within the proposed *Wise-ShopFloor* framework. The ToolZ (z_c), Pitch (θ_x), and Roll (θ_y) are the three independent variables that define the position and orientations of the moving platform of the PKM.

10.3 Conclusions

An integrated virtual environment for PKM design, analyze, validation, path planning, and remote control is proposed in the article, it can be used in the early stage for conceptual design of PKM and embodiment design stage with the CAD model and simulation. An example is implemented under the system. It is shown that the system is very efficient and generic for most of the PKM design and analyze.

References

1. Merlet JP (2000) Parallel robots. Kluwer, New York
2. Ares J, Brazales A, Busturia JM (2001) Tuning and validation of the motion platform washout filter parameters for a driving simulator. In: Driving simulation conference, pp 295–304
3. Gosselin CM, Guillot M (1991) The synthesis of manipulators with prescribed workspace. 123 ASME J Mech Des 113:451–455
4. Michalewicz Z (1994) Genetic algorithms + data structures = evolution programs. AI Series. 223 Springer, New York
5. Holland JH (1975) Adaptation in natural and artificial systems. The University of Michigan Press, Ann Arbor, MI
6. Abbasi WA, Ridgeway SC, Adsit PD, Crane CD, Duffy J (1997) Investigation of a special 6-6 parallel platform for contour milling. In: Proceedings of the ASME – manufacturing science and engineering division, pp 373–380
7. Akima T, Tarao S, Uchiyama M (1999) Hybrid micro-gravity simulator consisting of a high-speed parallel robot. In: Proceedings of IEEE International conference on robotics and automation, pp 901–906
8. Alizade R, Bayram C (2004) Structural synthesis of parallel manipulators. Mech Mach Theory 39:857–870
9. Alizade RI, Tagiyev NR (1994) A forward and reverse displacement analysis of a 6-dof in-parallel manipulator. Mech Mach Theory 29(1):115–124
10. Angeles J (1997) Fundamentals of robotic mechanical systems: theory, methods, and algorithms. Springer, New York
11. Ares J, Brazales A, Busturia JM (2001) Tuning and validation of the motion platform washout filter parameters for a driving simulator. In: Driving simulation conference, pp 295–304
12. Aronson RB (1997) Hexapods: Hot or ho hum? Manuf Eng 10:60–67
13. Arsenault M, Gosselin CM (2006) Kinematic and static analysis of a planar modular 2-dof tensegrity mechanism. In: Proceedings of the 2006 IEEE international conference on robotics and automation, pp 4193–4198
14. Asada H, Cro Granito JA (1985) Kinematics and statics characterization of wrist joints and their optimal design. In: Proceedings of the IEEE international conference on robotics and automation, pp 244–250
15. Bailey P (1994) The merits of hexapods for robotics applications. In: Conference on next steps for industrial robotics, pp 11–16
16. Baker JE (1992) On mobility and relative freedoms in multiloop linkages and structures. Mech Mach Theory 16(6):583–597
17. Barrilleaux J (2001) 3D user interfaces with java 3D. Manning Publications, Greenwich, CT
18. Behi F (1988) Kinematic analysis for a six-degree-of-freedom 3-prps parallel mechanism. IEEE J Robot Autom 4(5):561–565
19. Bergamaschi PR et al (2006) Design and optimization of 3r manipulators using the workspace features. Appl Math Comput 172:439–463
20. Bi ZM, Lang SYT (2009) Joint workspace of parallel kinematic machines. Robot Comput Integr Manuf 25(1):57–63

21. Bi ZM, Lang SYT, Zhang D, Orban PE, Verner M (2007) An integrated design toolbox for tripod-based parallel kinematic machines. Trans ASME J Mech Des 129:799–807
22. Bianchi G, Fassi I, Molinari-Tosatti L (2000) A virtual prototyping environment for pkm analysis and design. In: Proceedings of year 2000 parallel kinematic machines international conference
23. Boeij J, Lomonova EA, Andre JA (2008) Optimization of contactless planar actuator with manipulator. IEEE T Magn 44(6):1118–1121
24. Boër CR, Molinari-Tosatti L, Smith KS (1999) Parallel kinematic machines. Springer, London
25. Boudreau R, Gosselin CM (1998) The synthesis of planar parallel manipulators with a genetic algorithm. In: Proceedings of the ASME design engineering technical conference, Atlanta, GA
26. Boudreau R, Turkkan N (1996) Solving the forward kinematics of parallel manipulators with a genetic algorithm. J Robot Syst 13(2):111–125
27. Bruzzone LE, Molfino R (2003) Special-purpose parallel robot for active suspension of ambulance stretchers. Int J Robot Autom 18(3):121–129
28. Carbone G, Ceccarelli M (2005) A serial–parallel robotic architecture for surgical tasks. Robotica 23(3):345–354
29. Ceccarelli M, Lanni C (2004) A multi-objective optimum design of general 3r manipulators for prescribed workspace limits. Mech Mach Theory 39:119–132
30. Cervantes-Sánchez JJ, Rendón-Sánchez JG (1999) A simplified approach for obtaining the workspace of a class of 2-dof planar parallel manipulators. Mech Mach Theory 34:1057–1073
31. Chablat D, Angeles J (2002) On the kinetostatic optimization of revolute-coupled planar manipulators. Mech Mach Theory 37:351–374
32. Chen IM (1994) Theory and application of modular reconfigurable robotic systems. PhD thesis, California Institute of Technology
33. Chen NX, Song SM (1992) Direct position analysis of the 4–6 stewart platform. Robot Spat Mech Mech Syst 45:75–80
34. Chi YL (1999) Systems and methods employing a rotary track for machining and manufacturing. WIPO Patent No. WO 99/38646
35. Chirikjian GS (1995) Hyper-redundant manipulator dynamics: a continuum approximation. Adv Robot 9(3):217–243
36. Cléroux L, Gosselin CM (1996) Modeling and identification of non-geometric parameters in semi-flexible serial robotic mechanisms. In: Proceedings of the ASME mechanisms conference, Irvine, CA
37. Cohen R (1992) Conceptual design of a modular robot. ASME J Mech Des 112:117–125
38. Company O, Pierrot F (2002) Modelling and design issues of a 3-axis parallel machine-tool. Mech Mach Theory 37(11):1325–1345
39. Company O, Pierrot F, Fauroux JC (2005) A method for modeling analytical stiffness of a lower mobility parallel manipulator. In: Proceedings of the 2005 IEEE international conference on robotics and automation, Barcelona, pp 3232–3237
40. Culpepper ML, Anderson G (2004) Design of a low-cost nano-manipulator which utilizes a monolithic, spatial compliant mechanism. J Precis Eng 28(4):469–482
41. Culpepper ML, Kartik MV, DiBiasio C (2005) Design of integrated mechanisms and exact constraint fixtures for micron-level repeatability and accuracy. J Precis Eng 29(1):65–80
42. Davidor Y (1991) Genetic algorithms and robotics: a heuristic strategy for optimization. World Scientific, Singapore
43. Davis L (1991) Handbook of genetic algorithms. Van Nostrand Reinhold, New York
44. Earl CF, Rooney J (1983) Some kinematic structures for robot manipulator designs. J Mech Transm Autom Des 105(1):15–22
45. Fassi I, Tosatti LM, Negri S, Bernardo GD, Bianchi G (1999) A concept for a computer aided configuration tool for parallel kinematic machines. In: ICAR'99, pp 563–567
46. Fedewa D, Mehrabi M, Kota S, Gopalakrishnan V (2000) Design of a parallel structure ficture for reconfigurable machining systems. In: Proceedings of the 2000 Japan–USA flexible automation conference, pp 216–221

47. Ferraresi C, Pastorelli S, Zhmud N (1995) Static and dynamic behavior of a high stiffness stewart platform-based force/torque sensor. J Robot Syst 12(10):883–893
48. Fichter EF (1986) A stewart platform-based manipulator: general theory and practical construction. Int J Robot Res 5(2):157–182
49. Fogel LJ, Owens AJ, Walsh MJ (1966) Artificial intelligence through simulated evolution. Wiley, New York
50. Gallardo-Alvarado J, Rico-Martnez JM, Alici G (2006) Kinematics and singularity analyses of a 4-dof parallel manipulator using screw theory. Mech Mach Theory 41:1048–1061
51. Gen M, Cheng R (1997) Genetic algorithms and engineering design. Wiley, New York
52. Golbabai A, Seifollahi S (2007) Radial basis function networks in the numerical solution of linear integro-differential equations. Appl Math Comput 188:427–432
53. Goldberg D (1989) Genetic algorithms in search, optimization and machine learning. Addison-Wesley, Reading, MA
54. Goldberg DE, Samtani MP (1986) Engineering optimization via genetic algorithm. In: Ninth conference on electronic computation, pp 471–482
55. Gopalakrishnan V, Kota S (1998) A parallely actuated work support module for reconfigurable machining systems. In: Proceedings of 1998 ASME design engineering technical conferences, pp 1–9
56. Gosselin CM (1988) Kinematic analysis,optimization and programming of parallel robotic manipulators. PhD thesis, McGill University
57. Gosselin CM (1990) Stiffness mapping for parallel manipulators. IEEE Trans Robot Autom 6(3):377–382
58. Gosselin CM, Angeles J (1988) The optimum kinematic design of a planar three-degree-of-freedom parallel manipulator. ASME J Mech Transm Autom Des 110(1):35–41
59. Gosselin CM, Angeles J (1990) Singularity analysis of closed-loop kinematic chains. IEEE Trans Robot Autom 6(3):281–290
60. Gosselin CM, Guillot M (1991) The synthesis of manipulators with prescribed workspace. ASME J Mech Des 113:451–455
61. Gosselin CM, Wang JG (1997) Singularity loci of planar parallel manipulators with revolute actuators. J Robot Auton Syst 21:377–398
62. Gosselin CM, Zhang D (1999) Stiffness analysis of parallel mechanisms using a lumped model. Technical report, Département de Génie Mécanique, Université Laval
63. Griffis M, Duffy J (1989) A forward displacement analysis of a class of stewart platform. J Robot Syst 6(6):703–720
64. Gupta S, Tiwari R, Nair SB (2007) Multi-objective design optimization of rolling bearings using genetic algorithms. Mech Mach Theory 42:1418–1443
65. Hafez M, Lichter MD, Dubowsky S (2003) Optimized binary modular reconfigurable robotic devices. IEEE Trans Mechatron 8(1):152–162
66. Hamlin GJ, Sanderson AC (1997) IEEE Robot Autom Mag 4(1):42–50
67. Harary F (1969) Graph theory. Addison-Wesley, Reading, MA
68. Heisel U (1999) Precision requirements of hexapod-machines and investigation results. In: Boër CR, Molinari-Tosatti L, Smith KS (eds) Parallel kinematic machines – theoretical aspects and industrial requirements. Springer, Berlin, pp 131–150
69. Holland JH (1975) Adaptation in natural and artificial systems. The University of Michigan Press, Ann Arbor, MI
70. Hollingum J (1997) Features: Hexapods to take over? Ind Robot 24:428–431
71. Honegger M, Codourey A, Burdet E (1997) Adaptive control of the hexaglide, a 6 dof parallel manipulator. In: Proceedings of the IEEE international conference on robotics and automation, Albuquerque, USA
72. Hong KS, Kim YM, Choi C, Shin K (1997) Inverse kinematics of a serial manipulator: kinematic redundancy and two approaches for closed-form solutions. In: Proceedings of IEEE international conference on robotics and automation, pp 780–785
73. Hostens I, Anthonis J, Ramon H (2005) New design for a 6 dof vibration simulator with improved reliability and performance. Mech Syst Signal Process 19(1):105–122

74. Huang JY, Gau CY (2002) A pc cluster high-fidelity mobile crane simulator. Tamkang J Sci Eng 5(1):7–20
75. Huang T, Li Z, Li M, Chetwynd D, Gosselin CM (2004) Conceptual design and dimensional synthesis of a novel 2-dof translational parallel robot for pick-and-place operations. ASME J Mech Des 126:449–455
76. Hunt KH (1978) Kinematic geometry of mechanisms. Clarendon Press, Oxford
77. Hunt KH (1983) Structural kinematics of in-parallel-actuated robot-arms. ASME J Mech Transm Autom Des 105(4):705–712
78. Jensen KA, Lusk CP, Howell LL (2006) An xyz micromanipulator with three translational degrees of freedom. Robotica 24(3):305–314
79. Jin Q, Yang TL (2004) Synthesis and analysis of a group of 3-degree-of-freedom partially decoupled parallel manipulators. ASME J Mech Des 126:301–306
80. Jin Q, Yang TL (2004) Theory for topology synthesis of parallel manipulators and its application to three-dimension-translation parallel manipulators. ASME J Mech Des 126:625–639
81. Joines J, Houck C (1994) On the use of non-stationary penalty functions to solve constrained optimization problems with genetic algorithms. In: 1994 IEEE international symposium evolutionary computation, Orlando, FL, pp 579–584
82. Jovane F, Negri SP, Fassi I, Tosatti LM (2002) Design issues for reconfigurable pkms. In: 3rd Chemnitz parallel kinematics seminar: development methods and application experience of parallel kinematics
83. Klein CA, Blaho BE (1987) Dexterity measures for the design and control of kinematically redundant manipulators. Int J Robot Res 6(2):72–83
84. Kohli D, Osvatic M (1993) Inverse kinematics of general 6r and 5r, p serial manipulators. Trans ASME J Mech Des 115(4):922–931
85. Kong XW, Gosselin CM (2004) Type synthesis of 3t1r 4-dof parallel manipulators based on screw theory. IEEE Trans Robot Autom 20:181–190
86. Koren Y (1983) Computer control of manufacturing systems. McGraw-Hill, New York
87. Koren Y, Jovane F, Heise U, Moriwaki T, Pritschow G, Ulsoy G, VanBrussel H (1999) Reconfigurable manufacturing systems. CIRP Ann 48(2):6–12
88. Koza JR (1991) Evolving a computer program to generate random numbers using the genetic programming paradigm. In: Proceedings of 4th international conference on genetic algorithms, pp 37–44
89. Kucuk S, Bingul Z (2005) Robot workspace optimization based on a novel local and global performance indices. In: IEEE ISIE, pp 20–23
90. Landers R, Min BK, Koren Y (2001) Reconfigurable machine tools. CIRP Ann 49(1): 269–274
91. Landers RG (2000) A new paradigm in machine tools: reconfigurable machine tools. In: Japan–USA symposium on flexible automation
92. Lauffer J et al (1996) Milling machine for the 21st century-goals, approach, characterization and modeling. In: Proceedings of SPIE – the international society for optical engineering smart structures and materials, pp 326–340
93. Lauffer JP, Hinnerichs TD, Kuo CP, Wada B, Ewaldz D, Winfough B, Shankar N (1996) Milling machine for the 21st century – goals, approach, characterization and modeling. In: Proceedings of SPIE – the international society for optical engineering smart structures and materials 1996: industrial and commercial applications of smart structures technologies, San Diego, vol 2721, pp 326–340
94. Lee HY, Reinholtz CF (1996) Inverse kinematics of serial-chain manipulators. Trans ASME J Mech Des 118(3):396–403
95. Lee S, Kim S (1993) Efficient inverse kinematics for serial connections of serial and parallel manipulators. In: Proceedings of the 1993 IEEE/RSJ international conference on intelligent robots and systems, pp 1635–1641
96. Lewis G (1996) Automation technology: from imagination to reality. Catalog of Giddings & Lewis
97. Li T, Payandeh S (2002) Design of spherical parallel mechanisms for application to laparoscopic surgery. Robotica 20(2):133–138

98. Li YM, Xu QS (2008) Stiffness analysis for a 3-puu parallel kinematic machine. Mech Mach Theory 43(2):186–200
99. Lin W, Crane III CD, Griffis M (1994) Closed-form forward displacement analyses of the 4–5 in-parallel platforms. ASME J Mech Des 116:47–53
100. Lu Y, Hu B (2007) Analyzing kinematics and solving active/constrained forces of a 3spu+upr parallel manipulator. Mech Mach Theory 42:1298–1313
101. M GC, Zhang D (2000) Kinetostatic modeling of n-dof parallel mechanisms with passive constraining leg and revolute actuators. In: Proceedings of year 2000 parallel kinematic machines international conference
102. Manocha D, Canny JF (1992) Efficient inverse kinematics for general serial manipulators. In: Proceedings of the 1992 Japan–USA symposium on flexible automation, pp 125–129
103. Masory O, Wang J (1995) Workspace evaluation of stewart platforms. Adv Robot 9(4): 443–461
104. Matar G (1997) Hexapod: application-led technology. Prototyping Technol Int 70–720pt
105. Mehrabi M, Ulsoy G, Koren Y (2000) Reconfigurable manufacturing systems: key to future manufacturing. J Intell Manuf 11(4):403–419
106. Merlet JP (2000) Parallel robots. Kluwer, New York
107. Michalewicz Z (1994) Genetic algorithms + data structures = evolution programs. AI Series. Springer, New York
108. Mitchell JH, Jacob R, Mika N (2006) Optimization of a spherical mechanism for a minimally invasive surgical robot: theoretical and experimental approaches. IEEE T Biomed Eng 53(7):1440–445
109. Moon YM, Kota S (1999) A methodology for automated design of reconfigurable machine tools. In: Proceedings of the 3rd CIRP international seminar on manufacturing systems, pp 297–303
110. Nearchou AC (1998) Solving the inverse kinematics problem of redundant robots operating in complex environments via a modified genetic algorithm. Mech Mach Theory 33(3):273–292
111. Nguyen CC et al (1993) Adaptive control of a stewart platform-based manipulator. J Robot Syst 10(5):657–687
112. Owen J (1999) Tomorrow's machines in paris. SME Manuf Eng 123(2):118–129
113. Parenti-Castelli V et al (2004) On the modeling of passive motion of the human knee joint by means of equivalent planar and spatial parallel mechanisms. Auton Robot 16(2):219–232
114. Park MK et al (2001) Development of the pnu vehicle driving simulator and its performance evaluation. In: Proceedings of IEEE international conference on robotics and automation, pp 2325–2330
115. Pedrajas NG, Martnez CH, Perez JM (2002) Multi-objective cooperative coevolution of artificial neural networks. Neural Netw 15:1259–1278
116. Pérez R et al (2004) A modularity framework for concurrent design of reconfigurable machine tools. Lect Notes Comput Sci 3190:87–95
117. Pierrot F, Shibukawa T (1999) From hexa to hexam. In: Boër CR, Molinari-Tosatti L, Smith KS (eds) Parallel kinematic machines – theoretical aspects and industrial requirements. Springer, New York, pp 357–364
118. Pond G, Carretero JA (2006) Formulating jacobian matrices for the dexterity analysis of parallel manipulators. Mech Mach Theory 41:1505–1519
119. Portman VT, Sandler BZ, Zahavi E (2000) Rigid 6×6 parallel platform for precision 3d micromanipulation: theory and design application. IEEE Trans Robot Autom 16(6):629–643
120. Powell NP, Whittingham BD, Gindy NNZ (1999) Parallel link mechanism machine tools: acceptance testing and performance analysis. In: Boër CR, Molinari-Tosatti L, Smith KS (eds) Parallel kinematic machines – theoretical aspects and industrial requirements. Springer, New York, pp 327–344
121. Pritschow G (1999) Research and development in the field of parallel kinematic systems in europe. In: Boër CR, Molinari-Tosatti L, Smith KS (eds) Parallel kinematic machines – theoretical aspects and industrial requirements. Springer, New York, pp 3–15
122. Pritschow G, Wurst KH (1997) Systematic design of hexapods and other parallel link systems. CIRP Ann Manuf Technol 46(1):291–295

123. Pritschow G, Wurst KH (1999) Systematic design of hexapods and other parallel link systems. CIRP Ann 46(1):291–295
124. Houck CR, Joines JA, Kay MG (1995) A genetic algorithm for function optimization: a matlab implementation. Technical report, North Carolina State University
125. Ranganath R et al (2004) A force-torque sensor based on a stewart platform in a near-singular configuration. Mech Mach Theory 39(9):971–998
126. Rechenberg L (1973) Evolutionsstrategie: Optimierung Technischer Systeme nach Prinzipien der Biologischen Evolution. Frommann-Holzboog, Stuttgart
127. Rehsteiner F, Neugebauer R, Spiewak S, Wieland F (1999) Putting parallel kinematics machines (pkm) to productive work. Ann CIRP 48(1):345–350
128. Rey L, Clavel R (1999) The delta parallel robot. In: Boër CR, Molinari-Tosatti L, Smith KS (eds) Parallel kinematic machines – theoretical aspects and industrial requirements. Springer, New York, pp 401–418
129. Romiti A, Sorli M (1992) Force and moment measurement on a robotic assembly hand. Sens Actuators A 32:531–538
130. Rosati G, Gallina P, Masiero S (2007) Design, implementation and clinical test of a wire-based robot for neuro rehabilitation. IEEE Trans Neural Syst Rehabil Eng 15(4):560–569
131. Rout BK, Mittal RK (2008) Parametric design optimization of 2-dof r-r planar manipulator: a design of experiment approach. Robot Comput Integr Manuf 24:239–248
132. Ryu SJ, Kim JW, Hwang JC, Park C, Cao HS, Lee K, Lee Y, Cornel U, Park FC, Kim J (1999) Eclipse: an overactuated parallel mechanism for rapid machining. In: Boër CR, Molinari-Tosatti L, Smith KS (eds) Parallel kinematic machines – theoretical aspects and industrial requirements. Springer, New York, pp 441–455
133. Sefrioui J, Gosselin CM (1993) Singularity analysis of representation of planar parallel manipulators. J Robot Auton Syst 10:209–224
134. Shoham M et al (2003) Bone-mounted miniature robot for surgical procedures: concept and clinical applications. IEEE Trans Robot Autom 19(5):893–901
135. Sorli M, Pastorelli S (1995) Six-axis reticulated structure force/torque sensor with adaptable performances. Mechatronics 5(6):585–601
136. Sowizral H, Rushforth K, Deering (2001) The java 3D API specification. Addison-Wesley, Reading, MA
137. Stewart D (1965) A platform with six degrees of freedom. In: Proceedings of the institution of mechanical engineers, vol 180, pp 371–378
138. Stock M, Miller K (2003) Optimal kinematic design of spatial parallel manipulators: application to linear delta robot. J Mech Des Trans ASME 125:292–301
139. Sujan VA, Dubowsky S (2004) Design of lightweight hyper-redundant deployable binary manipulator. ASME J Mech Des 126(1):29–39
140. Tahmasebi F, Tsai LW (1994) Six-degree-of-freedom parallel "minimanipulator" with three inextensible limbs. United States Patent, Patent Number: 5,279,176
141. Takasaki S, Kawamura Y (2007) Using radial basis function networks and significance testing to select effective sirna sequences. Comput Stat Data Anal 51:6476–6487
142. Thomson WT (1993) Theory of vibration with applications. Prentice Hall, Upper Saddle River, NJ
143. Timoshenko SP, Gere JM (1972) Mechanics of materials. PWS Publication, Boston, MA
144. Tönshoff HK, Grendel H, Kaak R (1999) Structure and characteristics of the hybrid manipulator georg v. In: Boër CR, Molinari-Tosatti L, Smith KS (eds) Parallel kinematic machines – theoretical aspects and industrial requirements. Springer, New York, pp 365–376
145. Tosatti LM, Bianchi G, Fassi I, Boër CR, Jovane F (1997) An integrated methodology for the design of parallel kinematic machines. Ann CIRP 46:341–345
146. Tsai LW, Lee JJ (1989) Kinematic analysis of tendon-driven robotics mechanisms using graph theory. ASME J Mech Transm Autom Des 111:59–65
147. Uchiyama M (1994) A 6 d.o.f. parallel robot hexa. Adv Robot: Int J Robot Soc Jpn 8(6):601
148. Valenti M (1995) Machine tools get smarter. ASME Mech Eng 117:70–75
149. Verner M, and Xi F, and Mechefske C (2005) Optimal Calibration of Parallel Kinematic Machines, ASME Journal of Mechanical Design, 127:62–69

150. Wang L, Wong B, Shen W, Lang S (2002) Java 3d enabled cyber workspace. Commun ACM 45:45–49
151. Wang QY, Zou H, Zhao MY, Li QM (1997) Design and kinematics of a parallel manipulator for manufacturing. Ann CIRP 46(1):297–300
152. Wang SM, Ehmann KF (2002) Error model and accuracy analysis of a six-dof stewart platform. ASME J Manuf Sci Eng 124(2):286–295
153. Wapler M et al (2003) A stewart platform for precision surgery. Trans Inst Meas Control 25(4):329–334
154. Warnecke HJ, Neugebauer R, Wieland F (1998) Development of hexapod based machine tool. Ann CIRP 47(1):337–340
155. Wavering AJ (1999) Parallel kinematic machine research at nist: past, present, and future. In: Boër CR, Molinari-Tosatti L, and Smith KS (eds) Parallel kinematic machines – theoretical aspects and industrial requirements. Springer, New York, pp 17–31
156. Weck M, Giesler M, Meylahn A, Staimer D (1999) Parallel kinematics: the importance of enabling technologies. In: Boër CR, Molinari-Tosatti L, Smith KS (eds) Parallel kinematic machines – theoretical aspects and industrial requirements. Springer, New York, pp 283–294
157. Winter G, Periaux, Galan M, Cuesta P (1995) Genetic algorithms in engineering and computer science. Wiley, Chichester
158. Wurst KH (1999) Linapod – machine tools as parallel link systems based on a modular design. In: Boër CR, Molinari-Tosatti L, Smith KS (eds) Parallel kinematic machines – theoretical aspects and industrial requirements. Springer, New York, pp 377–394
159. Xi F, Zhang D, Xu Z, Mechefske C (2003) Comparative study of tripod-type machine tools. Int J Mach Tools Manuf 43(7):721–730
160. Xi F, Zhang D, Mechefske C, Lang SYT (2004) Global kinetostatic modelling of tripod-based parallel kinematic machines. Mech Mach Theory 39(4):357–377
161. Xu LJ, Fan SW, Li H (2001) Analytical model method for dynamics of n-celled tetrahedron variable geometry truss manipulators. Mech Mach Theory 36(11–12):1271–1279
162. Xu QS, Li YM (2006) A novel design of a 3-prc compliant parallel micromanipulator for nanomanipulation. Robotica 24(4):521–528
163. Yi BJ et al (2003) Design and experiment of a 3-dof parallel micromechanism utilizing flexure hinges. IEEE Trans Robot Autom 19(4):604–612
164. Yigit AS, Ulsoy AG (2000) Design of vibration isolation systems for reconfigurable precision equipment. In: Japan–USA symposium on flexible automation
165. Yoshikawa T (1984) Analysis and control of robot manipulators with redundancy. In: Proceedings of first international symposium on robotics research, pp 735–747
166. Yu A, Bonev IA, Paul ZM (2008) Geometric approach to the accuracy analysis of a class of 3-dof planar parallel robots. Mech Mach Theory 43(3):364–375
167. Yu X et al (2001) Measuring data based non-linear error modeling for parallel machine-tool. In: IEEE international conference on robotics and automation, pp 3535–3540
168. Zatarain M, Lejardi E, Egana F (1998) Modular synthesis of machine tools. CIRP Ann 47(1):333–336
169. Zhang CD, Song SM (1992) Forward position analysis of nearly general stewart platform. Robot Spat Mech Mech Syst 45:81–87
170. Zhang D (2000) Kinetostatic analysis and optimization of parallel and hybrid architectures for machine tools. Laval University, Canada
171. Zhang D, Gosselin CM (2001) Kinetostatic modelling of n-dof parallel mechanisms with prismatic actuators. Trans ASME J Mech Des 123(3):375–381
172. Zhang D, Mechsfske C, Xi F (2001) Optimization of reconfigurable parallel mechanisms with revolute actuators. In: CIRP 1st international conference on reconfigurable manufacturing
173. Zhang D, Mechsfske C, Xi F (2001) Stiffness analysis of reconfigurable parallel mechanisms with prismatic actuators. In: CIRP 1st international conference on reconfigurable manufacturing
174. Zhang D, Han W, Lang S (2003) On the kinetostatic analysis and dynamic modeling of a 3dof parallel kinematic machine. In: CIRP 2nd international conference on agile, reconfigurable manufacturing

175. Zhang D, Xi F, Mechefske C, Lang SYT (2004) Analysis of parallel kinematic machines with kinetostatic modelling method. Robot Comput Integr Manuf 20(2):151–165
176. Zhang D, Liu K, Ding F, Wong F (2005) Application of a parallel kinematic machine system in automation of light deburring operations. Trans CSME 29(2):229–245
177. Zhang D, Wang LH, Lang SYT (2005) Parallel kinematic machines: design, analysis and simulation in an integrated virtual environment. Trans ASME J Mech Des 127(7):580–588
178. Zhang D, Wang LH, Esmailzadeh E (2006) Pkm capabilities and applications exploration in a collaborative virtual environment. Robot Comput Integr Manuf 22:384–395
179. Zhang GL, Fu Y, Yang RQ (2002) Kinematics analysis of a new 3 dof planar redundant parallel manipulator. Mech Des Res 18:19–21
180. Zhang WJ (1994) An integrated environment for CADCAM of mechanical systems. PhD thesis, Delft University of Technology
181. Zhang WJ, Li Q (1999) On a new approach to mechanism topology identification. ASME J Mech Des 121:57–64
182. Zhao JS, Zhang SL, Dong JX, Feng ZJ, Zhou K (2007) Optimizing the kinematic chains for a spatial parallel manipulator via searching the desired dexterous workspace. Robot Comput Integr Manuf 23:38–46

Index

E

F

G

H

I

Breinigsville, PA USA
16 April 2010
236276BV00006B/10/P

9 781441 911162